PRAISE FOR *THE SECRET TEACHINGS OF PLANTS*

"I learned more from part one of this magnificent book than from any source in years. Buhner writes of complex discoveries in neuroscience and neurocardiology with clarity and coherence. Encompassing the highest spiritual insights of such giants as Blake, Goethe, and Whitman, part two is worthy poetry in itself, offering readers a unique way to move into transcendent realms. Of the truly great books appearing today, *The Secret Teachings of Plants* is easily the most rewarding I have had the privilege of reading."

JOSEPH CHILTON PEARCE, AUTHOR OF
THE BIOLOGY OF TRANSCENDENCE

"In this wonderful book Stephen Buhner shows us that the heart is not a machine but the informed, intelligent core of our emotional, spiritual, and perceptual universe. Through the heart we can perceive the living spirit that diffuses through the green world that is our natural home. Required reading for all owners of a heart."

MATTHEW WOOD, HERBALIST AND AUTHOR OF
THE BOOK OF HERBAL WISDOM

"Beautifully written, *The Secret Teachings of Plants* is a work of art—as much a poetical journey into the essence of plants as it is a guidebook on how to use plant medicine in our healing practices. Stephen Buhner is among the plant geniuses of our time. Like Thoreau and Goethe and Luther Burbank, the master gardeners and "green men" he so liberally quotes throughout, Buhner will be long remembered for his deep and introspective connection with the green world and for his ability to connect us to the heart of the plants through his teachings."

ROSEMARY GLADSTAR, AUTHOR OF *ROSEMARY GLADSTAR'S FAMILY
HERBAL* AND FOUNDER OF UNITED PLANT SAVERS

"Buhner's writings are a powerful call for people to work together to restore the sacredness of Earth."

BROOKE MEDICINE EAGLE, AUTHOR OF
BUFFALO WOMAN COMES SINGING

THE
SECRET TEACHINGS
OF PLANTS

THE INTELLIGENCE OF THE HEART IN
THE DIRECT PERCEPTION OF NATURE

STEPHEN HARROD BUHNER

BEAR & COMPANY
Rochester, Vermont

Bear & Company
One Park Street
Rochester, Vermont 05767
www.InnerTraditions.com

Bear & Company is a division of Inner Traditions International

Library of Congress Cataloging-in-Publication Data
Buhner, Stephen Harrod.
   The secret teachings of plants : the intelligence of the heart in the direct
perception of nature / Stephen Harrod Buhner.
      p. cm.
Includes bibliographical references and index.

   ISBN 978-1-59143-035-3 (pbk.)
1. Nature, Healing power of. 2. Heart. 3. Materia medica, Vegetable. I. Title.

RZ440.B84 2004
615.5'35—dc22

                                                                    2004019491

Printed and bound in the United States at Lake Book Manufacturing, Inc.

11

Text design and layout by Jonathan Desautels
This book was typeset in Sabon, with Bauer Text Initials as a display typeface

For Trishuwa
who completed the education of my heart

Where does the power come from
to see the race to its end?
From within.

CONTACT INFORMATION

More information on Stephen Buhner's work
and his apprenticeships and workshops
can be found at www.gaianstudies.org.

ACKNOWLEDGMENTS

Trishuwa who endured; Don Babineau who asked that it be written; Kate Gilday who asked that it be taught; Kathleen Maier who hosted the first weekends; Dale Pendell whose work birthed the form; Robert Bly for permission to use his translations of Goethe, Machado, Mirabai, Jiminez, and Baudelaire and especially his metaphor of the long bag we drag behind us; Benoit Mandelbrot for seeing the world with a child's eyes, Henri Bortoft for giraffes and insight; Henri Corbin for expressing the inexpressible; Rosita Arvigo and Matthew Wood for living a life expressed out of the reality of plant intelligence; James Hillman for his teachings of the heart; and Goethe, Henry David Thoreau, Luther Burbank, and Masanobu Fukuoka whose writings and lives stand testament to the intelligence of Nature.

Thanks are gratefully extended to the following publishers and authors for permission to reprint:

from *Living with Barbarians: A Few Plant Poems* by Dale Pendell, copyright 1999. Published by Wild Ginger Press, Sebastapol, California. Used by permission of the author.

from *The Taste of Wild Water: Poems and Stories Found While Walking in Woods* by Stephen Harrod Buhner, copyright 2003. Published by Raven Press, Randolph, Vermont. Used by permission of the author.

from *The Kabir Book* by Robert Bly, copyright 1971, 1977 Robert Bly, copyright 1977 The Seventies Press. Reprinted by Permission of Beacon Press, Boston.

CONTENTS

A NOTE TO THE READER

The first half of this book is linear, the second half is not. The first half is filled with analytical explanations of why and how—the second half is filled with poetry and doing. The first half of the book is called *systole* the second *diastole* to reflect the different natures of these two halves. The terms are usually used to describe the functioning of the heart. Systole is when the heart contracts, forcing blood outward away from the heart. Diastole is when the heart relaxes and fills once more. This book reflects that pattern—the movement away from the heart and the relaxation and movement inward as the heart fills again. In a sense this is one of the oldest patterns we know. Still, we have spent a long time as a culture in the systolic and you may find you aren't in the mood for more of it. So don't read the first half of this book if you don't want to—it is there if you want explanations later.

Feel free to skip around and read this book in any order you wish, choosing whichever chapter interests you and leaving those that do not. It doesn't matter, for the things that you need to find you will find, if only you will follow your heart.

I have long seen that each grain of knowledge I acquired, going to school to Nature, was added to each other grain I possessed, that these grains grew into a foundation stone, that the stones accumulated until I had a substructure, and that on that substructure I could build me a house. And I have seen, too, that there are enough buildings in Nature's system of knowledge to make a great city of wisdom.

I will never see that city completed; no man will. At best he may be able to construct, during his lifetime, one or two buildings, and perhaps to catch a vision of one or two streets and squares and parks and precincts of the whole. But the sublimity of the city—its endless boulevards, its imposing monuments, its transcendent capitol, its towering edifices, its vistas and sweeping panoramas—these we can only imagine, for the view we get of the structures of knowledge we ourselves are able to build up, grain by grain, rock by rock, tier by tier, story by story, through diligence and hard work, into one or two of the buildings we know are all there, somewhere, to be builded. When I think of this, I wonder why some men are content to erect nothing more than rude huts of knowledge—a little cabin of selfish learning, enough to house them while they amass money or gain power or win fame—and will not even try to raise some nobler structure of the wisdom Nature offers so freely and generously, and that any who come to her may have for the asking!

— LUTHER BURBANK

INTRODUCTION

The significant problems we face today cannot be solved at the same level of thinking we were at when we created them.

— ALBERT EINSTEIN

All the technical information was stolen from reliable sources and I am happy to stand behind it.

— EDWARD ABBEY

W E IN THE WEST HAVE BEEN IMMERSED in a particular mode of cognition the past hundred years, a mode defined by its linearity, its tendency to reductionism, and its insistence on the mechanical nature of Nature. This mode of cognition, the verbal/intellectual/analytical, is now the dominant one in Western culture. But it is becoming increasingly obvious that there are inherent problems with this mode of cognition and the assumptions about Nature that it possesses. As William James put it in *The Will to Believe*, "Round about the accredited and orderly facts of every science there ever floats a sort of dust-cloud of exceptional observations, of occurrences minute and irregular and seldom met with, which it always proves more easy to ignore than to attend to. The ideal of every science is that of a closed and completed system of truth [and] phenomena unclassifiable within the system are paradoxical absurdities, and must be held untrue."[1]

In our time, this dust-cloud of exceptional observations has become a whirlwind of powerful proportions. The primary mode of cognition that the practitioners of science have used during the past century—analytical, linear, reductionistic, deterministic, mechanical—has begun to reach the limits of its assumptions. For the particular mode of cognition used by scientists, and the system to which that mode has given rise, can only maintain coherence by leaving out or ignoring a great many events that did and do not fit within the neat system it created. The wild oscillations that are now occurring in

Nature, from global warming to uncontrollable forest fires, are an aspect of the consequences of that ignoring; we have begun to reap the whirlwind.

There is, however, another mode of cognition, one our species has used as our primary mode during the majority of our time on this planet. This can be termed the holistic/intuitive/depth mode of cognition. Its expression can be seen in how ancient and indigenous peoples gathered their knowledge about the world in which they lived, for example, and in how they gathered knowledge of the uses of plants as medicines.

All ancient and indigenous peoples said that they learned the uses of plants as medicines from the plants themselves. They insisted that they did not rely on the analytical capacities of the brain for this nor use the technique of trial and error. Instead, they said that it was from the heart of the world, from the plants themselves, that this knowledge came. For, they insisted, the plants can speak to human beings if only human beings will listen and respond to them in the proper state of mind.

Although these assertions have been disregarded by Western thinkers the past two hundred years—deemed the superstitious ramblings of unsophisticated, unchristian, and unscientific peoples—it is distinctly odd that every indigenous and ancient culture on Earth, cultures geographically and temporally distinct, would say the same thing. Surely, all the people who ever lived cannot be so similarly foolish as to have projected exactly the same kind of wishful or superstitious thinking onto the world. Surely, the people who have lived during the past two hundred years, and especially the past century, cannot have suddenly become so wise and intelligent that only they can understand the true nature of reality. All the billions of people who lived before them cannot have been so profoundly wrong.

There is tremendous hubris—and dangerous environmental perturbations—in disregarding the wisdom of the ancestors who have gone before us, people who said that they learned about the world not from the ability of their minds to work as analytical, organic computers, but from their hearts as organs of perception.

This more ancient mode of cognition has not disappeared just because another mode of cognition has gained dominance. The truth is that this capacity to learn directly from the world and plants has never been limited to ancient and indigenous cultures, even if the craft is now uncommon. It was used by the great German poet Goethe in the early nineteenth century in his discovery of the metamorphosis of plants, by Luther Burbank in the early twentieth century in his creation of the majority of food plants that we now take for granted. It was used by George Washington Carver in his work with and development of the peanut as a food, and is used now by Masanobu Fukuoka, the great Japanese farmer, in growing crops that consistently exceed the yields of farmers who use more scientific approaches. It was used by Henry David Thoreau, who was a great deal more than a naturalist, and even by Barbara McClintock, who won the Nobel prize for her work with transposons and corn genetics. The truth is that this way of gathering knowledge is inherent in the way we are structured as human beings. It is as natural to us as the beating of our hearts. It is not by nature a vague or mushy cognition, as reductionists so often assert. It is extremely elegant, sophisticated, and exact. The understandings that can be gained through this ancient mode of cognition exceed anything that what we now call *science* can or will discover or articulate about who and what human beings are or the world of which they are a part.

This gathering of knowledge directly from the wildness of the world is called *biognosis*—meaning "knowledge from life"—and, because it is an aspect of our humanness inherent in our physical bodies, it is something that everyone has the capacity to develop. It is, in fact, something that all of us use (at least minimally) without awareness in our day-to-day lives.

This ancient mode of cognition is crucially important for us, as a species, to reclaim, for we live in dangerous times. The threats to ourselves and the planet that is our home have never been more dire. These threats come from ways of thinking that are not sustainable, that bear little relation to the real world, and that are an inevitable error inherent in the linear fanaticism and *mechanomorphism* (seeing the world as a machine) of contemporary perspectives. They are threats that come from the dominance of one particular mode of cognition to the exclusion of all others.

To correct this imbalance, we need to come to our senses, to reclaim the ability each and every one of us has to see and understand the world around us (an ability that has been built into us over evolutionary time) in ways far more sustainable and sophisticated than reductionistic science can ever attain.

In this book I will tell you how this ancient way of information gathering occurs and how it can be used, both generally and in specific. It can be applied to anything: from the discovery of the medicinal uses of plants to understanding the living reality of a damaged organ system, from farming to the interrelationship of mycelial fungi and trees, from the intelligence of whales to the interconnected functioning of ecosystems.

But this mode of cognition is much more than a method for gathering more accurate and sustainable information about the world. Ultimately, it is a way of being, just as the linear mode of cognition is now (regrettably) a way of being. And, as a way of being, it is intimately concerned with things other than the mere extraction of knowledge from the heart of the world. It is concerned with our interconnection to the web of life that surrounds us. It is concerned with wholeness, rather than a focus on parts. It is concerned with the very human journey in which we are all engaged. It is intimately concerned with who we are and who we are meant to be in our time here in this life. For we are more than anything else an expression of the livingness of this world, and we were all born for a reason. The reconnection of our selves to the ground of being from which we have come, from which our species has been expressed over evolutionary time, opens up to us dimensions of experience that are essential in order for us to become ourselves.

But to understand how it is possible to gather knowledge from the heart of the world, without the dominance of the analytical mind or reductionistic, trial-and-error processes, it is crucial to first understand two things: that Nature is not linear and that the heart is an organ of perception.

SYSTOLE

OF NATURE AND THE HEART

The colors of the Dark One have penetrated Mira's body;
    other colors washed out.
Making love and eating little—those are my
    pearls and my carnelians.
Chanting beads and the forehead streak—those are my bracelets.
That's enough feminine wiles for me. My teacher taught me this.
Approve me or disapprove me; I praise the Mountain Energy
    night and day.
I take the path that ecstatic human beings have taken
    for centuries.
I don't steal money or hit anyone; what will you charge me with?
I have felt the swaying of the elephant's shoulders. . .
    and now you want me to climb on a jackass? Try to be serious!

— MIRABAI

PROLOGUE TO PART ONE

How much of life have I wasted by believing the thing I was taught, that thinking is what makes us better, that the brain is superior to heart.
— Author's Journal, June 2001

Like so many others in this century I found myself a displaced person shortly after birth and have been looking half my life for a place to take my stand. Now that I have found it, I must defend it.
— Edward Abbey

I REMEMBER THE FIRST TIME I heard the sound of my great-grandfather's heart.

I was born months early and the doctors put me into an isolet— a protective, enclosed crib. I was not often touched, or held, or breast-fed. And there I remained two weeks, touched only to be cleaned or bottle-fed, on a schedule.

At the end of that time, my family came for me. They took me to my grandmother's house, where the relatives had gathered. I remember the moment my great-grandfather took me in his hands and held me against his chest. The warmth of his hands, the scratchy–smooth feel of his starched white shirt. Then the smells: of starch, and his body, and the cigarettes he smoked. I remember, too, the sound of his breathing, his slow, soft inhale and exhale, and under it all, much deeper, the muffled reverberation of his heart.

Those interweaving sounds called to me, a symphony of breath and heart, washing over me like waters lapping an island shore. His every inhale and exhale pulling me, I moving to their ebb and flow. Their rhythms tugging, loosening me, and the shore being left behind. Currents taking me onward, into waters I had never known. My outflowing breath

6

his inhale, his exhale becoming my life. My heart absorbing his rhythms, two beats moving as one.

My tiny life was held in the embrace of his older and more powerful waves. And those waves were a language, carrying within them a meaning far older than words, telling me of being wanted and a part of something that would always be. Murmuring that in this place was my place, in this heart my heart. But deeper still, under all of that, there was a substance, some soul food that I needed to become human, that came to me in that moment of unity. I breathed it in with every breath, took it in with every heartbeat. A food as important to my spirit as my mother's milk to my body. And something in me, some tiny doorway, opened, and through it flowed this substance, this exchange of soul essence. Out of me, too, it flowed and he took it in in turn and his spirit rejoiced.

And without this bonding, this joining of two living beings, what is life? What is life without this exchange of soul essence but tasteless food in some dusty and empty place. And what are we then but abandoned and crumpled newspapers, yesterday's stories blowing down some wind-swept, darkened street.

I would visit my great-grandparents sometimes in the summers, traveling to their farm deep in rural Indiana. My great-grandfather and I would take walks in the woods, and sometimes, while we fished, I would lie close to him on the banks of the pond he had dug. I would smell his smell coming into me then and, as I settled deeper into this woodland place, I would notice again that soft inhale and exhale, feel once more the pull of ancient waters. And that soul force would flow into me, I would breathe it in like life itself. It seemed as if, as we lay there, the water, and the plants, and the trees over us, the very Earth itself, partook of that sharing. As if they, too, knew of this thing and blessed it and smiled upon us.

He died when I was eleven, and three years later my family moved to Texas. We lived in a house in a new subdivision where houses and streets were carved in geometric precision out of the Texas prairie. A mathematical model of community life, created in some university-trained architect's office, forced into place by bulldozer and concrete and Man, overlaying the drifting textures of the land.

I would go to the edges of the subdivision sometimes, when the workmen were gone, to where the new houses were rising, and I would enter them.

The smell of new wood,
disturbed sawdust
glittering in the sun.
Plywood flat on the floors,
and the empty echoing of my footsteps.

I remember those sights and sounds and smells, but mostly, it is the feelings in those places that I have never forgotten. There was something sad in those houses, something empty and forlorn. And I began to see, the longer I lived in that subdivision, that those same feelings were also to be found in the faces of my neighbors. There was a strange bewilderment, as if some part of these people was saying, "We have everything we are supposed to have to be happy; why then do we feel so empty and bereft?"

I would sometimes go beyond those rising houses into the corn fields that lay next to the geometric streets and houses. They also lay in ordered rows, a different kind of architecture forced onto the land. Sometimes I would disappear into those fields, tall corn closing over me, a world in each furrow and line. But sometimes I would walk even farther, into the woods that lay beyond. They were ragged and disordered woods, not like the forest I had known on my great-grandfather's farm. There were the marks of ax and bulldozer, plants crushed into tread marks, short stumps of large trees amid the small reminders of forest that had been allowed to remain.

The breath cannot be taken in deeply in such places, it rises shallow and short in the chest. The heart's beating is racing and rapid then, its thunder muted and soft. Like a tiny bird seeking release, fluttering desperately inside the chest.

Within such diminished landscapes I came into puberty. I knew that in time my face would become like my neighbors' faces. That some part of me would die there and that strange bewilderment would also emerge from my eyes. So, I filed emancipation papers and left home, the memory of what I had shared with my great-grandfather calling me, pulling me from the [safety] of that shore into waters I had never known. And sometime later, in the wildness of San Francisco in 1969, I met an interesting man.

Larry's red hair stood up, a rusty saw, its jagged spikes defying brush and gravity. His beard, like some strange reflection in a lake of still water, mirror-imaged his rusty and ragged head. When he spoke, his energy filled his face; his eyes widened, the whites standing out stark and distinct. In those moments, his hands took any excuse to burst forth and fly, grasping and moving the air to emphasize his meanings. I looked into his eyes as he talked and saw strange lands and people out of ancient legends, and his hands were callused and strong. And his stories were like none I had ever heard before.

After graduating high school, he had moved to mountains and built a cabin. He lived in seclusion for a year, eating little and talking less. Then he began to sail ships, tall masted schooners and small racing boats, and he sailed around the world. The typhoon hit him near the coast of Madagascar. Man working with cloth and rope and wood in a configuration as old as history while the heavens raged and the seas swelled and the ship was finally smashed against the coast.

Then one day, hearing him talk, something in my heart turned over, and a strange feeling went through me. I knew in that moment that there was something in mountains for me as well, something I had to find.

It is in such offhand ways that fate finds us and sets us on our way.

The first time I drove my car into the Rocky Mountains, those great peaks towered over me, soaring higher than I could see. The road snaked between them, following the winding of a river that had flowed at their feet for longer than the pyramids had touched the sky. Those peaks, like strange, wild, non-geometric office buildings, standing sentinel along the road, shadow-dimmed the light at the bottom of the canyon. I moved from shadow to light and back again, following the road that fate had given me. I kept on, and as I drove, moved farther from civilization, back into time, back into the wildness of the world.

And I was often afraid, for the road passes through areas where on one side the mountains tower up far into the heights and on the other drop into depths whose ending I could not fathom. Sometimes there are no guardrails and I could not help imagining what would happen if, for some reason, I left the safety of the road and lost control and soared out and down and down and down. So I held on tightly and the road took me ever deeper and farther on.

In time, I came to a place where the road crested, and I saw that if I continued on, it would begin to drop again, deep into the valleys, and take me back to the lowlands, to people, and the geometry of civilization.

There is something in us that sometimes drives us to the top, and we must go as high as we can and as far as the road takes us. So, far beyond tree lines and places where people lived, I found a turnout and stopped. I remember the sound the stones made as the car ground slowly to a halt, the slam of the door, the smell of hot engine and burning oil, and my steps on the gravel as I stood for the first time twelve thousand feet above the sea.

I paused then and was aware of the silence, as deep a silence as I have ever known. I could feel my heart beating within me and the coursing of my blood and could hear the slow, subtle, susurrus of my breathing. Then, suddenly, the power of the place flowed through my senses. The size of the mountains came into me and I felt their weight and then ever more deeply their age, and I was tiny and small and aware of something that had been, long before human beings had been, and would still be long after they were gone.

There was a trail off to my right that wandered through wildflowers, and stones, and short, matted gorse. Here and there great rock outcroppings arose, the craggy bows of wild ships whose decks I walked, and to keep my balance I was forced to lean into their pitch, like a sailor staggering on the wildness of the sea. The air was thin and cold and carried a smell that entered into me and has never left, and no amount of *city* will erase it nor length of time dull the remembering.

The trail wandered upward, the beaten ground showing there were those who had gone before me. I followed, and from time to time came upon streams that burst out of the ground and rushed down the mountain slopes, eager to join the river far below. When I put my hands in and lifted them to taste the wildness of that water, they were numbed to the bone; the water swirled through my mouth, a liquid ice. I could feel it twisting and turning as it moved deep within me. Something came into me with it, some wildness that city water no longer knows.

The trail wandered into a cleft between those ragged outcroppings of mountain, then opened out into a small protected glade. The great stones circled around, cupping the hollow within their palms, protecting it from the wind. A camp robber, ubiquitous bird of the heights, flitted close, landing on the circle of stones overlooking me. He cocked his head in question and called out to me something that was almost familiar. It was long before I knew what he had said.

A ragged, misshapen stone of granite nestled in the center of the glade, its sides covered with lichens orange and green. I bent down and could feel its textured surface, crumbly and warm beneath my hand. I sat then and leaned against it and felt the sun's warmth play over my face, the rock's surface rough against my back. There was a smell in the warm sunlight that day that I have never been able to describe, as if warm sunlight has a smell all its own. Other smells came to me as well, off the rock behind me, from the grass and wildflowers under and around me, from the air itself. And I felt the tension begin to go out of my body. I began to breathe deeply, held in the hands of the secret place I had found.

I started to hear the little sounds of the place then—the creak of rock, one side in sun, the other in shade. The soft fluttering of the wind. The wind leaned down and touched the plants, wildflowers and green stems of grass bending slightly under its caress. I felt its soft touch move to my face, its fingers following the line of my cheek, curving under my chin, ruffling my hair. In its slow, soft murmur the sigh and breathing of the world, flowing in and around me, like waters over a shore.

Their every movement pulling me, I moving with their ebb and flow. Their rhythms tugging, loosening me, and the shore being left behind. Currents taking me onward, into waters I had never known. My outflowing breath the world's inhale, its exhale now my life.

Then, slowly, my heart began beating with the rhythms of the glade, my tiny life held in the embrace of its older and more powerful waves. And those waves were a language, carrying within them a meaning far older than words, telling me of being wanted, a part of something that would always be. Murmuring that in this place was my place, in this heart my heart. But deeper still, under all of that, there was a substance, some soul food that I needed to become human, that came to me now. I breathed it in with every breath, took it in with every heartbeat. A food as important to my spirit as my mother's milk had been to my body. And something in me opened up, some tiny doorway within me, and through it flowed this substance. From me, too, it flowed, and the glade took it in and rejoiced. And in that moment I bonded with the world, as I had bonded with my great-grandfather so long before.

And without this bonding, what is life? What is life without this exchange of soul essence between the human and the wildness of the world? Tasteless food in some dusty and empty place rising in geometric precision out of an empty plain. A mathematical life forced into place by bulldozer and concrete and Man. And what are we then but abandoned and crumpled newspapers, stories without meaning, blowing down some wind-swept, darkened street.

So I found the thing I had sought, that had come into me first with my great-grandfather's heart. My eyes were soft focused, the colors of the land luminous, its sounds a rippling harmony of the rhythmic patterns of the world. I found my place.

In time I rose from that glade, found the trail and walked on, eventually coming to the crest. I stood and looked then and my eyes swept out and down, my vision soaring birds on currents of light. They traveled farther than I thought it possible for sight to go, gently touching the great foldings of those mountains, soaring from their valleys to their peaks. Then the wind blew against me and suddenly, for no reason that I knew, I was laughing, a wild, deep joy flooding me. The wind took my laughter in its hands and carried it up and out, into the wildness of the world.

I turned my eyes farther then, and across the valleys, far away, I could see a ragged wall of rain falling from darkened clouds high above, bending down to touch the Earth. On one side there was sun, the other darkened rain. A curtain of gray lace, hanging heavy from dark, water-laden clouds, was sweeping toward me, twisting and bending in the wind. Then some strange movement of clouds opened a path and the sun took it, laying light across that gray, lacy curtain. A rainbow spread out below me, cupping in its colors a brilliant blue lake whose wind-disturbed surface followed the ragged twistings and turnings of the land.

I felt my spirit move and then some *Mountain* thing touched me, looked down from a mighty height, awakened from its contemplation, to see me, tiny, below. And it was old beyond knowing and has little to do with humans and its gaze rocked me as, for a second, I stood revealed. Then it returned to a contemplation of centuries, of millennia, living a kind of life that is as far beyond me as the stars are from the sun.

Not long after that I returned to the car and drove on, eventually coming to a place lower down, where people go. There I found an old cabin and rebuilt it, and lived there, and began to make relationship with the wildness of the world. And from time to time, I would travel higher into the mountains and hike their forests.

I hunted in the high country for mushrooms and wild plants and followed the tracks of mountain men and the Indians who were there before them. I entered into the wild places and listened to their songs, for the world was young then; I was new and life stretched, unbroken, before me.

And though I had been taught in school that the wildness of the world was cold and uncaring, unfeeling, and ruled by tooth and claw, I did not find it so. It gave me all that I have ever wanted to have and began to teach me a truth that I had not learned in school, a truth plain in its every line, and movement, and turning. For Nature does not know how to lie.

It is such a simple observation that there are no straight lines in Nature. But it is a door into Nature's heart.

SECTION ONE

NATURE

When geometric diagrams and digits
Are no longer the keys to living things,
When people who go about singing or kissing
Know deeper things than the great scholars,
When society is returned once more
To unimprisoned life, and to the universe,
And when light and darkness mate
Once more and make something entirely transparent,
And people see in poems and fairy tales
The true history of the world,
Then our entire twisted nature will turn
And run when a single secret word is spoken.

— NOVALIS

CHAPTER ONE

THE NONLINEARITY OF NATURE

Deep in the human unconscious is a pervasive need for a logical universe that makes sense. But the real universe is always one step beyond logic.

— FRANK HERBERT

It is becoming obvious that channeled vision is not good enough. There must be a return from overspecialization to the generalist who can see totalities.

— CHANDLER BROOKS

Most men only care for science so far as they get a living by it, and that they worship even error when it affords them a subsistence.

— GOETHE

We shall see but little way if we are required to understand what we see. How few things can a man measure with the tape of his understanding! How many greater things might he be seeing in the meanwhile?

— HENRY DAVID THOREAU

I WENT TO COLLEGE, LIKE MANY PEOPLE IN THE 1970s, to escape the draft for the Vietnam war. When I first enrolled I didn't know what I wanted to study—I wasn't there to *learn*—and so I studied anything that seemed interesting to me. I wandered into philosophy, then the humanities, eventually washing up on mathematical shores. Though my learning seemed somewhat random to me then, it was not. I was searching for *explanations,* something to explain the deep experiences I had had, experiences that were profoundly important to me. For in my cultural myths I could find few traces of them.

Of course (though I did not know it then), the soul of the world cannot be found in philosophy, nor in the humanities, nor in mathematics, (nor even in science). It resides someplace else, someplace the linear mind cannot go. But to many people adrift, mathematics makes lover's promises and offers a safe harbor from the storm. "Here," it says, "are not only explanations, but a promise of total control." The rules are straightforward and understandable; unpredictability vanishes.

What about pi?

And those few irritating things, like pi, as the mathematicians discovered, can simply be rounded off. ("Ve haf vays of making you behafe.") Mathematics is almost always a profession for control freaks. It has little to do with life, very little to do with the real world.

Mathematics cannot eliminate prejudice, prevent willfulness, or resolve partisan differences. It has no power over anything in the moral realm.

— GOETHE

Anyone who truly looks will quickly see that Euclidean space is not present in a mountain range (nor is topology, but that is a different story); there are no straight lines, no rectangles, no spheres, no geometric angles of predictable value. Though this observation is a simple thing, obvious to any four-year-old, Western culture has ignored it for centuries, developing a culture expressed out of assumptions of Euclidean predictability. But life is not linear, its forms are not predictable to the linear mind, and it bears little relation to the mathematical reality developed by Euclid, the mathematics taught to all of us in school as geometry.

How hard it is to see what is right in front of us.

The word *geometry* is derived from the Greek *geometria: geo* meaning Earth and *metria* meaning to measure, literally, "to measure the Earth." But the term, has been corrupted, and now applies not to measuring the Earth, but to something that is not geometry at all—the measurement of Euclidean space. This may seem a ridiculous point to make, but our whole culture is based on the illusion that Euclid created with his mathematics. That illusion, which we take to be so very real, actually has little to do with the real world and nothing at all to do with natural environments like mountains, and oceans, and the places where water touches the land—coast lines. For Coast lines and Euclidean lines really have nothing in common. Coast lines have a very specific lack of smoothness, which makes them more complicated than any line Euclid ever imagined.

COAST LINES

When we are shown a map of land that is bounded on all sides by water, as islands such as Madagascar are, we also inevitably see its coastline. To get the area in square miles of such an island, geographers measure the coastline, calculate the distances across the land from coast to coast, and tell us that Madagascar has an area of 226,658 square miles. This is an application of Euclidean geometry to the world. But it is not real.

When calculating a measurement for a coastline, using Euclidean geometry, the living reality of a coastline is altered significantly. To understand why this has so little to do with the real world, you must remember just how a coastline *is* in the real world and not on maps. It is important to let the reality of its *being* enter your personal experience, to, perhaps, begin to see it once more with a child's eyes. For when you have done this, it becomes obvious just how little the geometry we are taught has to do with the real world in which we live.

> *It is a rare qualification to be able. . . to conceive and suffer the truth to pass through us living and intact. . . First of all a man must see, before he can say. . . See not with the eye of science, which is barren, nor of youthful poetry, which is impotent. . . . As you see, so at length you will say.*
>
> — HENRY DAVID THOREAU

When you approach a coastline, what you encounter is a ragged edge, some portions of which protrude farther into the water, some less. To measure this ragged line, Euclidean geometricians "round off" the raggedness. In essence, they take an approximation of the raggedness of the coastline in order to allow the complexity of a living coastline to fit into Euclidean space so that their model, their way of thinking, will be able to measure it. But always, it is important to remember, this is only an approximation. It is never real.

One way of understanding how coastlines are measured is to imagine that the line geographers measure is a path you can walk that goes around the whole island along the coast. Such paths rarely follow the exact edge of the coast; they are a bit inland from it, and their twistings and turnings are much less severe than the coastline itself. Walking the exact outline of a coast, with each little twist and turn, would be very hard to do. But if the path is a bit inland, as it is to make walking easier, that has more to do with comfort and ease than the coastline itself. So, to make the coast line more exact, you can imagine moving this path closer and closer to the water's edge, so that it follows more and more closely the irregular outline of the coast. And you can see that the closer the path gets to the water, the more wiggly it becomes. The more wiggly it becomes, the more twists and turns you must make as you walk it. The more twists and turns you make, the more you have to walk, and the longer the distance you must travel.

And so the closer the path to the exact edge where water and land meet, the longer the path will be. At some point in time, though, if the path keeps getting closer and closer to the water, you will be too big to follow all the ever-more-exact twists and turns along the coast, and so imagine now that the walker is a little mouse instead of you, a human being. A mouse can get much closer to the edge of the water than you can and thus follow all the twists and turns more easily. Doing this, of course, will make the path a lot longer for the mouse than it was for you because there are a great many more twists and turns, all of which have to be traveled, all of which have to be measured. This will work fine for a while, but if you keep moving the path closer and closer to the exact edge of the water, even the mouse will eventually be too big to follow all the tiny twists and turns. So, to follow the edge of the water more closely we must find a smaller walker, perhaps an ant. An ant, because of its size, can get even closer, following the twists and turns of the tiny pieces of the coastline ever more exactly. And again, the more closely the ant's path follows the exact outline of the coast, the more twists and turns there will be. So the length of the coastline becomes still longer as the walker becomes smaller. Eventually, even an ant will be too big to follow all the twists and turns and so perhaps we will have to imagine a microbe now traveling along the coast. Its size enables it to follow the coastline better still, and the line becomes, again, much more twisty and turny, and much, much longer.

All experiments show that with ever closer inspections, the mathematicians, "straight" lines become obviously ever less straight.
— BUCKMINSTER FULLER

Changing the size of the walker in the imagination is merely one way of increasing the magnification of the coastline. The smaller the point of view, the larger and longer the coastline becomes. Because the magnification can always be increased, the coastline can always become longer. Another way of saying this is that the length of a coastline grows the closer you get to the water's edge and the more closely you follow the coastline itself. The greater the scale of your measurement, the more twists and turns you must make to follow the coastline and the longer the line you are measuring becomes. (It can also be seen by this that the coast *line*—the path the walker takes—gets thinner and thinner the more exactly you approach the place where water and land meet, the more you increase your magnification. For an atom walking the coast, the line is very long and thin indeed.)

I find myself inspecting little granules, as it were, on the bark of trees, little shields or apothecia springing from a thallus, such is the mood of my mind, and I call it studying lichens. That is merely the prospect which is afforded me. It is short commons and innutritious. Surely I might take wider views. The habit of looking at things microscopically, as the lichens on the trees and rocks, really prevents my seeing aught else in a walk. Would it not be noble to study the shield of the sun on the thallus of the sky, cerulean, which scatters its infinite sporules of light through the universe? To the lichenist is not the shield (or rather the apothecium) of a lichen disproportionately large compared with the universe? The minute apothecium of the pertusaria, which the woodchopper never detected, occupies so large a space in my eye at present as to shut out a great part of the world.
— HENRY DAVID THOREAU

Nature is like this. If you take any part of Nature and look at it, its edges will be ragged and twisty and not straight. So to measure parts of Nature, to make Nature accessible to linear—some would call it "practical"—thinking, all of us are taught to round off the edges, to create specific and distinct boundaries between one thing and the next. This in actuality only gives us an approximation, a guess. And no matter the power of our magnification, it remains only an estimation of the real. For a long time, scientists ignored this (and this ignoring has had terrible consequences). They spent their time seeing Nature through Euclid's eyes and Newton's (who thought this way, too) and ignored the simplest thing there is about Nature, something any child can see immediately: it goes on and on and on.

I have never known a clergyman or a professor who could be more narrow, bigoted, and intolerant than some scientists, or pseudo-scientists. . . Intolerance is a closed mind. Bigotry is an exaltation of authorities. Narrowness is ignorance unwilling to be taught. And one of the outstanding truths I have learned in my University [of Nature] is that the moment you reach a final conclusion on anything, set that conclusion up as a fact to which nothing can be added and from which nothing can be taken away, and refuse to listen to any new evidence, you have reached an intellectual dead-center, and nothing will start the engine again short of a charge of dynamite. . . Ossified knowledge is a dead-weight to the world, and it does not matter in what realm of man's intellectual activities it is found. . . Any obstinate clinging to outworn doctrines, whether of religion or politics or morality or of science, are equally damning and equally damnable.

— LUTHER BURBANK

So the idea of quantitatively measuring Nature, of measuring a coast-line, of "coastline length" itself, as the mathematician Benoit Mandelbrot remarks, "turns out to be an elusive notion that slips between the fingers of one who wants to grasp it. All measurement methods ultimately lead to the conclusion that the typical coastline's length is very large and so ill determined that it is best considered infinite. Hence . . . length is an inadequate concept."[1]

In nature, a whole encloses the parts, and yet a larger whole encloses the whole enclosing the parts. By enlarging our field of view, what is thought of as a whole becomes, in fact, nothing more than one part of a larger whole. Yet another whole encloses this whole in a concentric series that continues on to infinity.

— MASANOBU FUKUOKA

Thus, the length of the coastlines listed in school books are never accurate, for, with increasing magnification, the actual coastline in the real world becomes longer and longer. In order to measure coastline length, geographers smooth out the irregularities that are inherent in Nature and in real coastlines. Only in this way can their mode of measuring be used. But this smoothing out ignores an essential facet of Nature—its nonlinearity. Coastlines, in actuality, in reality, continually approach infinite length, and any assumption that they are measurably finite forces a nonlinear system into a linear mode of cognition. And this always leaves out something, and that something is very important indeed.

THE SUBJECTIVITY OF SCIENCE

Any measurement of Nature that smooths out its irregularities in order to allow measurement is not objective. It is, in fact, highly subjective.

The observer, by determining the degree of measurement (or magnification) that will be used, and thus how the lines will be smoothed out, interferes with what is being measured. The observer intervenes in any resultant description of Nature by subtly altering its description, a description that depends on a preference for one level of magnification over another. It is an error that is not rectifiable—not correctable—because the error comes from the way of thinking itself. It comes from applying a linear, static mode of cognition to a nonlinear, always changing and flowing reality. That this resultant description is then taken as an accurate portrayal of Nature injects an unreality into our collective consciousness. We are slightly moved away from Nature, and everything we do begins to take on perturbations that grow greater the farther away in time we go from, and the more decisions we make based upon, that original error in description.

The truth is that in the real world, in Nature, quantification is *always* a projection of arbitrary human decisions. It is always subjective. Nature contains no fixed, measurable quantities.

What about those four rocks over there?

We often think, because we are so thoroughly immersed in Euclid's imaginary world in school and by our culture, that there are quantities in Nature. We are shown a number of oranges and we think there is a quantity—seven, perhaps. But number and quantity are not the same thing. As Gregory Bateson admonishes, "You can have exactly three tomatoes. You can never have exactly three gallons of water. Always quantity is approximate."[2]

Nature may contain numbers of things, but it contains no quantities of things—only *qualities*.

I'm starting to feel sick

And when we are taught, and come to believe, that rigorous thinking, that scientific thinking, occurs only with the exactitude of quantification, we embark on a course that is more a reflection of the kind of thinking we are engaged in (and unconscious, unexamined projections) than it is the real world. It has almost nothing to do with the real world or Nature. It is, in fact, crazy.

> *Scholars have for the most part a diseased way of looking at the world. They mean by it a few cities and unfortunate assemblies of men and women, who might all be concealed in the grass of the prairies. . . When I go abroad from under this shingle or slate roof, I find several things which they have not considered.*
> — HENRY DAVID THOREAU

As children, of course, we instinctively understand the nearly infinite, open-endedness of Nature. It is only with schooling that we lose our natural understanding (and lack of fear) of the nonlinearity of the world. If you ask young children to guess the length of any particular coastline they will joyfully do so. But when you explain to them about mice and ants and microbes they will realize immediately that any coastline is as long as you want to make it, and they will laugh, knowing that Nature goes on and on and on. For they experience this truth every day of their lives when they play in the secret worlds they find in their yards.

(Coastlines—like all things in Nature—are finite in length, even though their limits can never exactly be found. Even though finite, they approach an infinite length. They are wrinkled for just this reason, in order to extend their length, to have them approach infinity as closely as possible, to have them go on and on and on.)

It is only adults, when given this exercise of thinking about coastlines, who will become afraid of what it means, who will feel the underpinnings of their reality begin to crumble, and refuse to accept it or its implications.

A finite living being partakes of infinity, or rather, has something infinite within itself. We might better say: in a finite living being the concepts of existence and totality elude our understanding; therefore we must say that it is infinite, just as we say that the vast whole containing all beings is infinite.

— GOETHE

The recognition that there was something wrong in applying linear thinking to life was recognized long ago, of course, and captured in language (inevitably) by a Greek named Zeno of Elea—though he put the paradox in story form as a race between Achilles and a tortoise. Zeno of Elea's paradox states that there is between me and that wall a distance that must be traveled before I can touch the wall. The length of that distance, the line between me and the wall, can be divided in half, and that half in half again, and that half in half again, and so on, forever. So it is never possible to reach that wall, for an infinite distance must be traveled to reach it.

Straight lines are axiomatically self-contradictory and self-cancelling hypothetical ventures.

— BUCKMINSTER FULLER

The internalization of this paradox, once it is grasped, is often frightening (nausea sometimes accompanies its ingestion), for it challenges the foundations of the linear thinking to which we have become habituated. Most people, once they grasp it and internalize it (and feel the fear that it engenders) dismiss and ignore it as senseless, as irrelevant. But it reveals a profound truth about the nature of linear thinking and its limitations. Obviously, I can reach the wall, and so something must be wrong (hence the paradox of the thing). The something that is wrong is not in the paradox itself, but in the thinking that gives rise to it. It has to do with applying linear thinking to life. Life is not, and never has been, linear.

(The roiling in the gut that can accompany the recognition of the non-linearity of Nature, of the absence of quantity in Nature, is an *experience* of the sickness, of the aberration, of linear thinking when it is used as a dominant window through which the world is seen. The repercussions of that thinking can be *seen* in the destruction of wild landscapes, in the logging of rain forests and the damning of rivers.)

Euclid defined the dimensions of physical matter that we now take for granted: a point has no dimensions, a line one, a rectangle two, and a sphere three. But if you think about it, you have never *seen* a no dimensional, one-dimensional, or two-dimensional form—they do not exist in Nature and they never have. They existed only in Euclid's mind and now, regrettably, in ours.

> *Planes are not experimentally demonstrable. Solids are not experimentally demonstrable.*
>
> — BUCKMINSTER FULLER

Each of these increasing dimensions, in Euclid's world, stands at a ninety degree angle to the one preceding it. We are used to thinking of physical objects through this matrix, this definition of the dimensions of matter. We are taught that we live in three-dimensional space. But Euclid limited his mathematics to (imaginary) objects for which all the dimensions coincided. Thus we are taught in school that shapes are regular and are mathematically measurable. We constantly apply this kind of thinking to all of Nature but what is true is that Nature is not regular and its shapes are not dimensionally concordant, not so regular and predictable. Nature's increasing dimensions do not necessarily stand at ninety degrees to the ones preceding them. (Hence, the nonlinearity—the *chaos*—of the thing.)

The shapes around which Euclid arranged his mathematics are exceptionally rare in Nature: mountains are not cones, the Earth is not a sphere, and straight lines are nonexistent.

> *Someday someone will write a pathology of experimental physics and bring to light all those swindles which subvert our reason, beguile our judgement and, what is worse, stand in the way of any practical progress. The phenomena must be freed once and for all from their grim torture chamber of empiricism, mechanism, and dogmatism.*
>
> — GOETHE

Any object in Nature will display the same kind of unsettling dynamics coastlines do when they are deeply examined. The supposed two-dimensionality of a rectangular plain and assumed three-dimensionality of a mountain will themselves approach infinite size as the magnification of measurement increases. (There is *no* linearly measurable length, width, or height. No quantity to them at all.) Euclid's world is not the real world, and his system of measurement only works accurately in his imaginary world. In Nature, something much different is going on, and trying to understand it with the linear mind gets complicated, for Nature is as far beyond lines as the stars are from the sun.

> *Science has made no experimental finding of any phenomena that can be described as solid, or as continuous, or as a straight surface plane, or as a straight line, or as infinite anything.*
> — BUCKMINSTER FULLER

The truth is that what Euclid left out of his mathematical world was *life*. When life flows through (what is called) three-dimensional space, it changes it. The smooth lines twist, fracture, wiggle, and fold over themselves in all directions. And this twisting and fracturing occurs not only along the dimensional line itself—along the one-, two-, or three-dimensional line being considered—but *between* the dimensional lines themselves. Thus, shapes in nature are composed of discordant dimensions. A mountain is not a cone or a pyramid that possesses three distinct and clear dimensions, each at ninety degrees to the other. As life flows through (physical) space, each dimensional line of a mountain fractures and folds, approaching infinity in length.

As well, the mountain's dimensionality itself is not a constant, it fluctuates somewhere between two and three—a fractional dimension—because as life flows through the so-called three dimensions, not only are the dimensional direction lines of height, width, and depth broken and fragmented, but the *space* through which they flow is also broken and fragmented. So this dimensionality of mountains is always spilling over into another dimension at different degrees of spill at different points along its near-infinite fractional lines. The greater the dimension of a non-linear object, such as a mountain, the greater the chance that a given region of space contains a piece of it. Thus, where the mountain begins and ends can never be determined. It seems to have a beginning and an end. But we will never know, with the linear mind, just where that end is or if, in fact, it actually exists.

The fact is, that you cannot see all of the facts about anything just by looking at the thing itself. To learn part of the essential truth about grasses, for instance, you have to study the cow!. . . A fact is relative, and if it is placed out of its relative position it apparently is not a fact, often.

— LUTHER BURBANK

We can see the ripples that life caused when it fell into the land in the upthrust jagged peaks that we call mountains. But upon what distant shore do those ripples end? Do they end on the sandy beach of the ocean? However tiny they are, however unseen they are to the linear mind, are those ripples not present still? And just as the shade of the oak tree is present in the seed, is not the eagle present in the mountain? And if the eagle flies to the field, is not the mountain now in the field? The waters begin in the mountain snows, but when they flow to the sea, does not part of the mountain reside now in the ocean?

Nonlinear structures—the shapes found in Nature—are the visible remnants of the movement of life through matter.

(Even this way of talking is too reductionistic. The question "What came first, the chicken or the egg?" is a product of the linear mind. Linearity is an illusion. *Life* came first, and within it all living forms are inherent.)

Each shape has its own particular identity and the linear mind names them mountains or coastlines or trees (though to name something is never to know it). We see them as static entities, as if we are outside them. To the linear mind they appear to be static and unchanging. But they are not.

You cannot get out of Universe. Universe is not a system. Universe is not a shape. Universe is a scenario. You are always in Universe. You can only get out of systems.

— BUCKMINSTER FULLER

When the linear mind looks at Nature or a part of Nature, it takes a picture of it, frozen in one instant of time. If the linear mind looks at a moving, changing process—the flight of a bird, for instance—it takes a series of pictures, one after the other. Each picture shows the presence of the bird in a different location at a different moment in time. But these strung-together snapshots are not the flight of the bird, however much they appear to be so to the linear mind. Even if the linear mind could move as quickly as a motion picture camera, it still would only be capturing static moments of the flight of the bird. Even film can never capture the flight of the bird, for a moving picture is only a series of snapshots moving very rapidly, giving the appearance of flight. No matter how fast the shutter speed of the camera, there will always be a tiny piece of time left out from all such series of pictures. The semblance of a living thing is found in this process, but only a semblance. The linear mind, and the moving picture cameras it invents, will always leave out that tiny piece of time. And it is in that tiny piece of time that the oh-so-hard to describe but oh-so-strongly felt thing we know as life resides. It is something that always resides between and outside the frozen moments that are graspable by the linear mind.

> *A collection of an infinite number of parts includes an infinite number of unknown parts. These may be represented as an infinite number of gaps, which prevent the whole from ever being completely reassembled.*
> — MASANOBU FUKUOKA

Scientists, when focusing their research on one particular thing, take a picture, with the linear mind, of the moments and the movements of that living thing through (supposed) three-dimensional space. They separate it out and isolate it. They take a piece of Nature, remove it from the flow of life and time, and study it, trying to understand Nature and life or, perhaps—to them—more simply, the leaf of a plant. But once it is removed from its living context, broken off from the matrix within which it exists, it is no longer what they think it to be. This unnatural separation can never produce the outcome they desire, and everything they decide based on that separation will always end up wrong.

> *One of our great limitations is our tendency to look only at the static picture, the one confrontation. We want one-picture answers; we want key pictures. But we are now discovering that they are not available.*
> — BUCKMINSTER FULLER

FRACTALS, NONLINEARITY, AND DETERMINISTIC CHAOS

So . . . any close observation of a natural object reveals a highly irregular structure. The greater the magnification of our vision, the more irregular the object's surface will become. And because nearly all natural formations on Earth are irregular, trillions upon trillions of them, they cannot be described using Euclidian geometry or the mathematics of Newton. This was recognized by Benoit Mandelbrot, who has long had the habit of seeing the world like a child and asking difficult questions. Because he could not find a word in any language (including mathematics) to describe the infinite, irregular shapes of Nature he invented one—fractal.

Mandelbrot created the term from the Latin word *fractus,* which means "something broken apart into irregular shapes." Fractus is also the origin of the English words *fraction* and *fragment.* And so a fractal is something that has irregular, nonperiodic shape. That is, it is fractional, fragmented. (And this use of the word fraction or fractal to describe Nature evokes the fundamental realization that everything we see, including ourselves, is only a fractional part of one very large whole.)

The fractal shape of natural objects means they are irregular or disjoint and the word stands in opposition to *algebra* which comes from the Arabic *al-jabr* and means "to rejoin broken parts." (It was used originally to refer to setting broken bones.) Euclidean geometry uses algebra to measure shapes and it joins together the nonlinearity of nature by smoothing out its irregularities into something that can be understood, and supposedly controlled and predicted, with the linear mind. But fractals are not Euclidean and they are intimately related to life itself. They are not a static system of three-dimensional shapes. Fractal lines—the fractal geometry of Nature—are the shapes created when life flows *through* physical space. And they are *always* in flux. Looking out from our tiny and limited life span, we continually miss the fact that life is *still* flowing through physical space. It has never stopped. The mountain lives much slower than we do, but its shape is *never* static, never unchanging. It is always flowing along and between dimensions, in constantly fluctuating, never predictable ways. This understanding disturbs deeply embedded (adult, not child) preconceptions and species bias, about matter and Nature, about what is living and what is not. For human cultures to allow scientists to dissect Nature as much as they have, Nature had to become a dead, unalive thing, otherwise no one would have put up with it.

> *When we allowed*
> *science to convince us*
> *that there is no soul*

or intelligence in matter,
the Earth's physical forms
became only cemetery markers
showing where spirits once moved
through the world.
The autopsy
of the material world
then began in earnest.
Its dissected parts
now litter the landscape
and we walk, depressed,
among lifeless statuary,
only accidental lifeforms
on the surface of
a ball of rock
hurtling around the sun.

The metal gate is unlocked.

Other kinds of flowers
nod in sunlight
outside that wrought-iron fence.

Recognizing the nonlinearity of Nature confounds the linear mind; to truly understand Nature we are forced to think outside the (appropriately Euclidean) box, to abandon quantity in favor of quality. For the linear mind, this removal of dimensional, quantitative, (living/nonliving), limitations means the removal of all points of (mental) reference. It is inherently frightening. As Mandelbrot comments, "Almost every case study we perform involves a divergence syndrome. That is, some quantity that is commonly expected to be positive and finite turns out either to be infinite or to vanish. At first blush, such misbehavior looks most bizarre and even terrifying, but a careful reexamination shows it to be quite acceptable. . . *as long as one is willing to use new methods of thought.*"[3]

A far more difficult task arises when a person's thirst for knowledge
kindles in him a desire to view nature's objects in their own right and

in relation to one another. . . he loses the yardstick which came to his aid when he looked at things from the human standpoint.

— GOETHE

The fractal nature of Nature, the nonlinearity of actual objects in the real world can be thought of as an added dimension to all natural forms. And this dimension must be taken into account when describing Nature. For if it is not, something other than Nature is being described. Describing Nature, naming a thing, is a wonderful yet perilous act. Once people have a name for something, their tendency is to think they understand it and once they think they understand it, they quit experiencing it fresh and new each time they encounter it. Should the name itself be inaccurate it starts a chain of cultural and individual events that lead to outcomes that are not predictable in the initial act of naming.

Semen *is Latin*
for a dormant, fertilized,
plant ovum—
a seed.
Men's ejaculate
is chemically more akin
to plant pollen.
See,
it is really
more accurate
to call it
mammal pollen.

To call it
semen
is to thrust
an insanity
deep inside our culture:
that men plow women
and plant their seed
when, in fact,
what they are doing
is pollinating
flowers.

Now.
Doesn't that change everything between us?

Ultimately life must be, is intended to be, experienced. Life is no mere description. To experience life, to get to the heart of things, to see truly the face of Nature—not just to describe it through the framework of an illusionary, disinterested, objective observer—a nonlinear mode of cognition must be used. For life is, as Frank Herbert realized, "always one step beyond logic."

CHAPTER TWO

THE SELF-ORGANIZATION OF LIFE

The mathematical intuition so developed [by training students in linear mathematics] ill equips the students to confront the bizarre behaviors exhibited by the simplest discrete nonlinear systems. . . yet such nonlinear systems are surely the rule, not the exceptions.

— R. M. MAY

There is in all things a pattern that is part of our universe. It has symmetry, elegance, and grace—those qualities you find always in that which the true artist captures. You can find it in the turning of the seasons, in the way sand trails along a ridge, in the branch clusters of the creosote bush or the patterns of its leaves.

— FRANK HERBERT

In the human spirit, as in the universe, nothing is higher or lower; everything has equal rights to a common center which manifests its hidden existence precisely through this harmonic relationship between every part and itself.

— GOETHE

IF YOU LOOK CLOSELY AT ANY LINE IN NATURE you will find a fractal line; it will be wrinkly and irregular. And if you look at a section of that wrinkled line through a microscope or under any greater magnification, this increased perspective will reveal yet smaller wrinkles on the larger ones. And further magnification of that smaller part will reveal yet smaller wrinkles. This all continues on and on for a very long time.

Each series of smaller wrinkles will be found to be very similar in shape to the larger one you started with (see figure 1.1). This is true of all natural objects such as branching trees, coral formations, wrinkly coastlines, ragged mountain ranges, the heart and circulatory system, plant leaves, and the brain and central nervous system. Thus a fractal, to be even more precise, is a nonlinear object composed of subunits (and sub-sub-units) that resemble the larger structure.

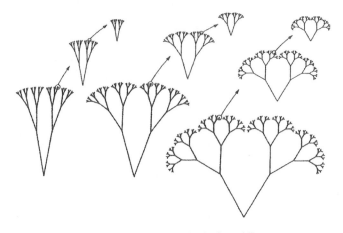

Figure 1.1. Self-similarity in fractal lines

This is the grandeur of Nature, that she is so simple, and that she always repeats her greatest phenomena on a small scale.
— GOETHE

All fractal objects possess this property, known as self-similarity; while they are highly irregular, they also have patterns. They are not simply chaotic, that is, unpatterned randomness. And even more difficult for the linear mind, this also applies to any processes or *properties* the objects may have, such as velocity, pressure, and temperature.

Every property of a natural object will, when examined, display a fractal nature. For example, the temperature of the human body is *never* an unchanging 98.6 degrees Fahrenheit. Examined on a graph, this always-changing body temperature will be a ragged and wrinkled line, just like a coastline. And if you take a small section of that wrinkled line and magnify it, it will possess the same kind of irregular pattern that the larger line possesses. And if you magnify a smaller section of that smaller line, it will also possess smaller wrinkles, all self-similar to the larger one. And so on and on and on. Just as quantity does not exist in space, neither does it exist in time. We may have the quality of warmth, but we never have a quantity of it.

Rather than fractal shapes, these are fractal *processes*. Instead of being fractal in space, they are fractal in time. Fractal processes generate irregular fluctuations on multiple time scales, just as fractal shapes generate irregular structures on multiple length scales. And it turns out that the fluctuations of these processes are quite often *oscillations*. They ebb and flow, increase and decrease in intensity, like a sound wave pattern or waves on the ocean.

There is a tendency, when recognizing that fractal patterns exist, to once again apply linearity of thought and assume that while the (supposed) lines and planes of nature are fractal, the pattern underneath is always predictable and regular. But this is, again, an inaccuracy. The oscillating patterns underneath are themselves expressions of nonlinearity. The patterns themselves express fractal dimensionality.

this is a hard habit to break

This sort of fractal, oscillatory pattern is extremely vivid when examining coastlines, for coastlines alter considerably with the tides. The movement of the moon and the gravity well that accompanies it pull the waters of the Earth along as the moon moves in orbit around the Earth. And so, in following the moon's pull, the waters ebb and flow. This ebb and flow is an oscillating movement, phase-locked to the moon's movements. Thus the coastline itself is a constantly shifting identity whose exact orientation in space and time is always fluctuating. There is a regularity about this oscillation, but it is not linear. Examining the oscillation of the tides and thus the shifting of any particular coastline will reveal that the oscillations are nonlinear, fractal processes. Any close examination of this nonlinearity will reveal that it contains smaller subunits and sub-subunits of oscillation that are all self-similar.

But this picture is still too reductionistic, too much a viewing of the world as a place of objects, things that have shape and sometimes movements—all with some mechanical basis. But there is nothing in the world that is merely mechanical, nothing that is not alive.

Virologists have been too busy, for instance, with their DNA-RNA genetic code isolatings, to find time to see the synergetic significance to society of the fact that they have found that no physical threshold does in fact exist between animate and inanimate.
— BUCKMINSTER FULLER

These apparently static material forms, mountains and water, are the body and blood of a living ecosystem, the Earth, and can never be accurately viewed in isolation from the whole. They make up one complete, living organism.

In pursuing investigations on the border region of physics and physiology, I was amazed to find boundary lines vanishing, and points of contact emerge between the realms of the Living and Non-Living. Metals are found to respond to stimuli; they are subject to fatigue, stimulated by certain drugs and "killed" by poisons.
— JAGADIS CHANDRA BOSE

We must step up the complexity of all of this to see it more clearly; the discomfort to the linear mind must be greater.

MOLECULAR SELF-ORGANIZATION

When a large number of molecules congregate in close proximity, the random motions of the billions and billions of molecules will at some point show a sudden alteration in behavior; all of them will start to spontaneously synchronize. They begin to move and vibrate together. They begin acting in concert, actively cooperating, and become tightly coupled together into one, interacting whole exhibiting a collective, macroscopically ordered state of being. They become a unique living system of which the smaller subunits (the molecules) are now only a part (Riding a bicycle is a very simple example of this. At the moment of balance, the two of you become a single, self-organized system.) During molecular synchronization, the molecules combine into a system that is *self-organized*. At such a moment, an entity has come into being; life has flowed through physical space. And the edges of this newly self-organized system are fractal in nature. Although the system is now organized, it is not linear, not a Euclidean form or system. Something new, nonlinear, has come into being.

> *Whatever appears in the world must divide if it is to appear at all.*
> *What has divided seeks itself again, can return to itself and reunite. . .*
> *in the reunion of the intensified halves it will produce a third thing,*
> *something new, higher, unexpected.*
>
> — GOETHE

We have started here with something already highly self-organized, of course: a molecule. It, too, is composed of smaller subunits and sub-sub-units, all of which are self-organized as well, all of which display fractal-ization. And like the linear mind taking snapshots of the flight of a bird, if any part, any subunit, is separated out from the whole and viewed in isolation, that tiny instant of time between one snapshot and the next is lost. And it is this thing, held in the tiniest instant of time, that is of the essence. Life will never be found in the DNA nor any *part* of the whole. Life is the thing that is more than the sum of the parts, the thing that hap-pens at the moment of self-organization, the nonlinear *quality* that comes into being at the moment of synchronicity.

In that moment of self-organization, the system begins to display something other than synchronicity as well. It begins to *act* as a unit, to have *behaviors*. The whole, tightly coupled system begins to act upon its microscopic parts to stimulate further, often much more complex, syn-chronizations. A continuous stream of information begins flowing back and forth, extremely rapidly, between the macroscopic, ordered whole to the smaller microscopic subunits and back again so that the self-organiz-ing structure is stabilized, its newly acquired dynamic equilibrium actively maintained. This information stream also immediately includes the exter-nal environment, where a similar rapid flow of information occurs, in order to more fully enhance stability. The system is now displaying *emer-gent* behaviors.

> *Self-organization initiates the fractalization of matter,*
> *emergent behaviors initiate the fractalization of time.*

Some of these behaviors will be simple things like temperature fluctuations, velocity, and pressure. Some of them are much more complex.

In self-organized systems, the information from the smaller subunit—which travels to the larger whole as chemical cues, electromagnetic fluxes, pressure waves, and so on—creates a response in the larger system, which is fed back to the initial site as a new informational pulse. This informational waveform travels through the system affecting and altering everything it touches. And these informational pulses travel back and forth, extremely rapidly, for as long as the system itself remains self-organized.

These pulses of information are fractal processes. They are composed of subunits and sub-subunits that are all self-similar to the larger informational pulse or process. They are intended to enhance the overall stability of the system, and they show the same kinds of nonlinear patterning present in all fractals. These emergent behaviors make up a complex controlled feedback system whose self-similar scaling property enhances the stability of the system. For not only at the very large macroscopic level, but also at the smaller subunit and still smaller sub-subunit and sub-subsubunit levels, these informational pulses are interacting, stabilizing the system from its tiniest microscopic subunit up through each succeeding subunit to the larger system itself, in a near-infinite cascade.

These kinds of spontaneously self-organized systems occur at many levels of complexity in living organisms. A system at one level of complexity—a molecule—can join together with others to form systems that self-organize at new levels of complexity, such as cells. When sufficient numbers of self-organized systems come together, all the self-organized groups will suddenly begin to synchronize and form one larger coherent unit. As groupings from each level join together to form new levels of complexity, entirely new types of self-organization occur, the form and behavior of which cannot be understood or predicted from any study of the preceding systems. To separate out and examine any one part of the whole in isolation—a cell or organ, for example—misses the point all over again. That tiny thing that occurs at the moment of self-organization is more than the sum of the parts.

Western man firmly believed nature to be an entity with an objective reality independent of human consciousness, an entity that man can know through observation, reductive analysis, and reconstruction. . . In his efforts to learn about nature, man has cut it up in little pieces.

He has certainly learned many things in this way, but what he has examined has not been nature itself.

— MASANOBU FUKUOKA

The harder the linear mind tries to grasp this reality, the more slippery it becomes. A self-organized system is a living, ever changing identity that comes into being of its own accord in a gesture of acquiescence and cooperation that is never static.

But these attempts at division also produce many adverse effects when carried to an extreme. To be sure, what is alive can be dissected into its component parts, but from these parts it will be impossible to restore it and bring it to life again.

— GOETHE

And while the excited (linear) mind may sometimes focus on the whole system—applying a hierarchy of importance to greater levels of complexity—the subunits themselves are neither more nor less important than the whole. Biological processes are the consequence of a dynamic, interactive, nonlinear network in which all parts play an equally important role. The system itself cannot exist, could not have come into being, without the subunits that self-organized. And the removal of too many of the subunits from a mistaken belief that they are unimportant will result in the loss of self-organization and emergent behaviors. In ecosystem studies, this is known as a *trophic cascade*. It occurs when too many parts of the ecosystem are destroyed and the nonlinear, self-organized ecosystem begins to collapse.

The things we call the parts in every living being are so inseparable from the whole that they may be understood only in and with the whole.

— GOETHE

The entire system and all its parts are cooperative, not competitive. They make up one system. They are whole.

OF CLOWNS AND UNICYCLES

At some never predictable point the increasing number of molecules crosses a *threshold,* beyond which occurs the moment of self-organization. On one side, there is nothing but randomized molecular movements, on the other side sudden self-organization and emergent behavior. (This threshold line itself is not a like a fence line; it is, in fact, much like a coastline, and like a coastline its exact orientation in space and time ebbs and flows.) At the moment the threshold is crossed, at the moment when self-organization occurs, the new living system enters a state of dynamic equilibrium. And to maintain self-organization, the system constantly works to maintain that state of dynamic equilibrium, much like a clown balancing on a unicycle.

Because a unicycle is not a static, stable perch like a chair, remaining balanced on it necessitates constantly adjusting the human–unicycle orientation in space and time. When a clown balances on a unicycle, he is always moving slightly this way and that in response to any perturbations that occur. There are always tiny factors that affect his balance and, instinctively, as he learns to ride the unicycle, the clown will, automatically, slightly adjust his balance responses in order to remain in equilibrium.

Thus, a clown sitting on top of a unicycle in one place is an example of a dynamic system engaging in constant change in response to alterations in its environment—slightly moving all the time in order to keep balance. If the clown becomes totally stationary, he and the unicycle will fall over. They will destabilize and no longer form a self-organized, whole system. They will fall apart, separate into parts: clown on one side, unicycle on the other.

The clown's motions are expressions of the precise corrections needed to stabilize an unstable perch. And those precise corrections occur in response to the *information* that is encoded in any perturbation that affects his dynamic equilibrium. Each perturbation is interpreted extremely rapidly. The information encoded within the perturbation tells the clown—at a level far below that of the conscious mind—exactly what that perturbation will do to his balance. His body understands the information and devises a complex, coordinated response of his entire being to maintain his equilibrium.

This is deterministic chaos—nonlinear dynamics. An intricate, underlying order exists—stabilization of the clown on the unicycle—but the actions that will occur for that stabilization to take place can never be predicted. The state the clown is in is extremely close to the threshold between balance and falling over, stability and the loss of stability. He constantly senses the alterations or perturbations—the *information*—that affect his balance, and he shifts his behavior in response.

All living organisms—all self-organized systems—are like this: They all retain an exquisite sensitivity to perturbations of the equilibrium that occurred when they self-organized. They *remember* that moment of equilibrium; they are attuned to it. The threshold itself is a living identity to them. They very closely monitor their internal and external world through extremely tight couplings, at billions upon billions of points of contact, in order to process the energy, matter, and information that is coming to them. These couplings occur in space through their nonlinear, fractal geometries and in time through their nonlinear, fractal processes.

Self-organized systems are living identities that engage in continual communication, both internal and external. They are not isolated, static units that can be understood in isolation. To examine them in isolation kills the living entity itself, and paying attention to the thing and not its communications—its balance-initiated information exchange—reveals very little about the true nature of what is being studied.

I have just been through the process of killing [a box-turtle] for the sake of science; but I cannot excuse myself for this murder, and see that such actions are inconsistent with the poetic perception, however they may serve science, and will affect the quality of my observations.
— HENRY DAVID THOREAU

That fractal geometry is found in the surfaces of self-organized systems is important, for it is actually a highly sophisticated and crucial aspect of maintaining stability. The folding and fracturing that occurs along and between dimensions in living organisms allows them to couple with—to touch—the world around them at a nearly infinite number of points, a great many more than if their edges were merely straight lines. For when any organism wrinkles its exterior (or any interior) surface, it tremendously increases the area of that surface and the length of its edges. This increase significantly expands the organism's ability to gather information from its external and internal environments. And when it wrinkles its *functioning*, it tremendously increases the number of possible behavioral responses available to it. Having a nearly infinite number of responses allows an organism to maximize its behavioral options for any potential internal or external environmental flux that its nearly infinite touching reveals to it. Since a self-organized system can never know just what future events might occur that will destabilize it, having a nearly infinite number of responses available increases its survivability immensely.

The fractal nature of living organisms allows a near infinite surface area, with resulting near infinite points of interaction, allowing maximum flexibility in response to environmental flux.

Thus, all self-organized, nonlinear systems exhibit an enormous range of behaviors in their ongoing work to maintain their equilibrium. The alteration of any one internal or external parameter causes the system to slightly move back over the threshold of self-organization, to go, momentarily, into *disequilibrium*. This forces it to nearly instantaneously alter its behaviors in order to reattain equilibrium, just like the clown on the unicycle.

These alterations are not predictable and can occur in the system's form or behavior or both. Each different bifurcation—each different pathway that is developed in response to an external or internal perturbation—leads to different expressions of form, behavior, and states of information storage and transfer. So, even if you have two systems that are identical when they self-organize, they will begin to diverge more and more from one another over time. For the perturbations that each of them experience will never be the same and can never be predicted, and each system's responses will be slightly different. At the moment of disequilibrium, each living system makes a *choice* from among the millions upon millions of available ways of reestablishing equilibrium. And that choice can never be predicted.

ah, free will

Over long periods of time, similar organisms will diverge so much that they will appear extremely different to the eye in form and function. What one sees then is the complex divergence of life that the Earth holds. There will not simply be plants that all look the same, but instead a multitude of forms that may appear so different as to seem unrelated.

Anything a self-organized, living system detects—anything that touches it—affects its balance. And this stimulates the system to shift its functioning, however minutely, in order to maintain its dynamic equilibrium. All nonlinear systems—all living organisms—are like this. And what facilitates their ability to respond to the minute touches of the world upon them is that they are not in a permanent equilibrium, not in a static state of being. They are poised, powerfully balanced, held in dynamic tension from one tiny fractal moment to the next. There is no one state to which they return when they are disturbed. They are always shifting, altering themselves, always about to fall into disequilibrium from environmental perturbations and always reorganizing—reestablishing a dynamic equilibrium—in new ways.

Thus, nonlinear systems can change in a sudden, discontinuous fashion, creating significantly new physical forms and behaviors in a very short period of time. Chemical production can alter so much in a single generation in plants living in different ecosystems that two plants identified as identical may bear little chemical relationship to one another.

As physician Ary Goldberger remarks, "For nonlinear systems, proportionality does not hold: small changes can have dramatic and unanticipated effects. An added complication is that nonlinear systems composed of multiple subunits cannot be understood by analyzing these components individually. This reductionist strategy fails because the components of a nonlinear network interact, i.e., they are coupled. Examples include the 'cross-talk' of pacemaker cells in the heart or neurons in the brain. Their nonlinear coupling generates behaviors that defy explanation using traditional (linear) models."[1]

> *I thoroughly understand that there are scientists to whom the world is merely the result of chemical forces or material electrons. I do not belong to this class.*
>
> — GEORGE WASHINGTON CARVER

The things that affect a living organism's functioning, and the responses it makes, cover a very broad range. Perturbations (and responses) can be chemical, mechanic, hormonal, electromagnetic, gravitic, and so on, in near-infinite variation and form. They can be either simple or complex, periodic, aperiodic, nonperiodic, or pulsatile, fast or slow, and include amplitude or frequency modulation.

Physicist Freidemann Kaiser notes, "The type of the external stimulus is irrelevant (mechanical, chemical, hormonal, electromagnetic, etc.). It is the information contained in the signal that is significant."[2]

Thus it is the information, the meaning encoded within the perturbation, that is important, not the form in which it is delivered. The form is merely one possible language of communication out of myriad possibilities. In the end it is the *meaning* inside the behavior that is significant, not the behavior itself. It is not the chemical released, nor the movement of the body, nor the electromagnetic field that is important, but the information, the meaning, that it carries.

And for too long, scientists have assumed that there is no meaning in Nature. As a result, they have spent their time studying static, dead forms, when the communications of meaning themselves are the essential thing. (It is no wonder then, that after years of schooling, so many of us now believe that life is meaningless, or that scientists have made Prozac to help us not notice how we feel.)

> *The grammarian is often one who can neither cry nor laugh, yet thinks that he can express human emotions.*
> — HENRY DAVID THOREAU

No matter how much you dissect the words and structure of the sentence you are reading now, such a study of its parts will never reveal its *meaning*. You may look at the history of language, how one word evolved into another or was combined with one from another continent through long interaction, study the function of verbs, adverbs (and their proliferation), nouns, adjectives, interjections, dangling participles, split infinitives, vowels and consonants, their shapes and sounds, proper pronunciation and articulation, but the meaning resides someplace else. It is embedded within the sentence, but it is not present in its parts. Those parts, in a sense, have "self-organized" to generate the meaning. The meaning is *not* the word, just as the territory is not the map.

There is a tension between the words, something joining them together, a pattern that emerges into consciousness that is not contained within any of the parts when they are considered separately.

All life is like this. There is a tension between the parts, something connecting them together, a pattern that emerges into consciousness that is not contained within any of the parts when they are considered separately. And this something that is not in the parts is the whole ball of wax, the point, the complete shebang.

> *Life as a whole expresses itself as a force that is not to be contained within any one part.*
> — GOETHE

This whole is almost always considered to be outside of the realm of science because it does not lend itself to reductionism. As a consequence, most scientists know nothing about it.

THE NONLINEAR DYNAMICS OF LIVING ORGANISMS

The self-organization that occurs when billions and billions of randomly fluctuating molecules suddenly synchronize is a hallmark of life and its emergence in myriad forms.

(This description of it is not *it,* of course. The most important thing is that any system that self-organizes can be *felt;* it possesses *qualities.* In that change from one moment to the next, something new comes into being, something never present in this world before and never to be repeated. A full life consists of encountering these millions upon millions of self-organized systems, *feeling* that extra something that comes into being at their emergence, *touching* it, *interacting* with it, *living* with it.)

Over time, such groupings of synchronized molecular systems fuse together as cells, the basic building blocks upon which all complex life rests. Cells, though tiny, are extremely intricate living systems that express self-organization and emergent behaviors. Like all living systems, they are extremely sensitive to external perturbations. The number of external perturbations they must detect and respond to is tremendously large, and many are extremely subtle. Researcher Adam Arkin relates, "The cellular program that governs cell cycle and cell development does so robustly in the face of a fluctuating environment and energy sources. It integrates numerous signals, chemical and otherwise, each of which contains, perhaps, incomplete information of events that the cell must track in order to determine which biochemical subroutines to bring on— and off-line, or slow down and speed up. These signals, which are derived from internal processes, other cells, and changes in the extracellular medium, arrive asynchronously and are multi-valued; that is, they are not merely 'on' or 'off' but have many values of meaning to the cellular apparatus. The cellular program also has a memory of signals that it has received in the past, and of its own particular history as written in the complement and concentrations of chemicals contained in the cell at any instant."[3]

All self-organized systems are, in fact, intelligent. They have to be. For they must continually monitor their environments, internal and external; detect perturbations; decide on the basis of those perturbations what the likely effect will be; and respond to them in order to maintain self-organization.

Man wants to believe that his intelligence and his ability to think and to have ideas removes him from what he always speaks of as "the

*lower orders;" he would much rather believe, as the old Greeks and
Romans did, that his leaders and great men were descended direct
from gods, and that he himself has natural privileges and preroga-
tives that are denied dogs. . . when I consider the use my dogs have
made of the opportunities that were theirs, I can't feel very vainglo-
rious about the superiority of man.*

— LUTHER BURBANK

Cells, like all self-organized systems, remain very close to the threshold between nonequilibrium and equilibrium in a kind of self-organized
criticality (SOC). Systems with SOC, like cells and avalanches, are close to
critical states. A signal (the vibration from an explosion or the noise made
from walking on snow) pushes them over the critical threshold and they
fall into disequilibrium. All the millions upon millions of signals or perturbations that impact cells affect their equilibrium. They process the
information encoded in the stimulus that pushed them back into disequilibrium and use it to generate behaviors that restore equilibrium.

So cells, and all self-organized systems, wobble continuously between
balance and imbalance. The self-organized systems we know as life, and
the behaviors that come from them, could not exist without the delicate
state of dynamic equilibrium that occurs between balance and imbalance.
Life comes from the constant interaction between chaos and order.
Without the dark, the light would have no meaning or purpose.

For any living system very close to the phase transition between a synchronized and nonsynchronized state, a small perturbing signal produces
a very large effect, moving the system in and out of synchrony at a regular rate. But every time such a system reorders, it is in a *new* state of equilibrium. The resulting self-organization and emergent behaviors are
different from those that went before. Thus novelty in living systems arises
at points of instability, at bifurcation points. Instabilities are indispensable
sources of biological innovation. Sometimes these instabilities lead to the
unique fusion of multiple self-organized systems into new organisms, as
the microbiological researcher Lynn Margulis has found in her work on
mitochondria. She calls this *symbiogenesis.*

*No living thing is unitary in nature; every such thing is a plurality.
Even the organism which appears to us as an individual exists as a
collection of independent living entities.*

— GOETHE

Mitochondria are the power generators for our cells, the intracellular power factories for our metabolism. But mitochondria are more than this. They are formerly free-living bacteria who were incorporated into cells long ago. Margulis has found their wild relatives still living as independent organisms, much as they did before this evolutionary fusion took place. What she discovered is that two kinds of cells came together and fused into one new organism, an organism that possesses capacities that the cells did not have before they fused together.

As with self-organized molecules, there was a point at which the two organisms crossed over a threshold. And at that crossing, they joined together into one self-organized system with new emergent behaviors. They began acting in concert with one another, actively cooperating, becoming a tightly coupled, interacting whole that exhibited a collective, macroscopically ordered state of being. Margulis's research showed that evolution is the emergence of individuality from the interblending of once-independent organisms; that evolutionary novelty arises from symbiosis or the merged, mutual collaboration of different self-organized systems.

As more and more molecular (or cellular) groupings—each with self-organization—come together, the more complex the living system becomes. (Still, no living system, however complex, can or will rival the complex, living matrix from which it was expressed—Nature itself.) The more interacting elements there are, the more sensitive the living system becomes to any perturbations affecting its dynamic equilibrium. And because the actions of such systems are nonlinear, the prediction of the behavior of these living systems cannot occur from a study of the component parts or be reduced to a study of single molecules and their impact on a system (as is done in standard medical and scientific research). For to do so, the living system must be viewed as static and unvarying, except in response to the introduction of the molecule being studied.

No living system is static and unvarying. They all exist in a state of dynamic equilibrium whose exact shape and behavior alters from millisecond to millisecond in response to external perturbations. And millions upon millions of perturbations are occurring at any one moment in time to all living organisms.

Nature, herself, has no hard and fast mode of procedure. She limits herself to no grooves. She travels to no set schedule. She proceeds an

inch at a time—or a league—moving forward, always, but into an
unmapped, uncharted, trackless future.

— LUTHER BURBANK

One of the important aspects of nonlinearity is that the sensing of living systems adapts very rapidly to (and quits noticing) regular, periodic
stimuli. But nonlinear stimuli are always new, never predictable and regular. This keeps the sensory noticing capacity of the system unhabituated,
so that it always notices perturbations. And the more sophisticated its
ability to notice perturbations, the more elegantly it can respond to them.
In fact, living systems have developed highly sensitive mechanisms to
notice perturbations that are extremely weak, whether they are chemical,
mechanical, or electromagnetic. The more sensitive these systems are to
even the weakest perturbation, the more they are able to enhance their stability. They can detect perturbations that are so subtle that scientists have
insisted they are too weak to exert effects. (Like the extremely tiny parts
per trillion of plant chemicals in the environment.)

In fact, the sensory systems of living organisms operate very near the
theoretical limits that can be calculated for sensing weak signals from a
noisy environment. For one of the abilities of systems with long-range
coherence is that they can detect much weaker signals than any individual component of that system. Further, they have the ability to enhance
such signals.

CHAPTER THREE

THE ENERGETICS OF LIFE

Biological cells can be viewed as highly sophisticated information-processing devices that can discern complex patterns of extracellular stimuli. In line with this view is the finding that, in analogy to electrical circuits, biochemical reaction networks can perform computational functions such as switching, amplification, hysterisis, or band-pass filtering of frequency information.

— JAN WALLECZEK

When in the exercise of his powers of observation man undertakes to confront the world of nature, he will at first experience a tremendous compulsion to bring what he finds there under his control. Before long, however, these objects will thrust themselves upon him with such force that he, in turn, must feel the obligation to acknowledge their power and pay homage to their effects.

— GOETHE

The scientist, unable to see light as anything other than a purely physical phenomenon, is blind to light.

— MASANOBU FUKUOKA

The electromagnetic spectrum has been around for a long time, much longer than human beings. Our use of the electromagnetic spectrum to broadcast radio and television signals is not really the innovation we have been led to believe. Life in all its forms has been using the electromagnetic spectrum for communication for billions of years.

> *The physical Universe is an aggregate of frequencies.*
> — Buckminster Fuller

All living organisms receive electromagnetic signals all the time. And like the signals received by our radios, many of them contain extremely large amounts of information, which can be used for a great many things. These range from regulating the opening of little doors in cells to let food in and waste out, to healing, to the beating of the heart, to birds orienting themselves to the magnetic lines of the Earth when migrating, to the communication between pollinators and their flowers, to the communications between members of the same family who have bonded with each other— and, of course, a great deal more.

Electromagnetic spectrum signals, like those we know as a particular radio station, can and do contain very large amounts of information. While driving in our cars we can receive an incredible range of information: from news of a flash flood on the road, to the complete contents of the *Encyclopedia Britannica,* to a song (which itself contains a great deal of information in the lyrics and the melody). If we happen to have a cell phone in the car, we can not only receive information, but can also broadcast it. If we have a CB radio, our broadcasts will go not just to one person, but to everyone who happens to be listening on that frequency.

While human beings are proud of the communications technology they have developed, they are, in fact, only Johnny-come-latelies. Life on Earth has been using the electromagnetic spectrum to send and receive signals filled with highly sophisticated information for nearly 4 billion years. The electromagnetic spectrum is simply another facet of Universe, another dimension through which life can flow. And when life flows through the electromagnetic spectrum, when it flows through a particular frequency, it fractalizes it, just as other dimensional lines fractalize. The oscillating sine wave or broadband frequency that life flows through becomes a fractal and its edges take on the same kind of irregular shaping that solid objects do.

Every time life flows through a frequency in the electromagnetic spectrum, it fractalizes that wave differently, because the flow of life is always nonlinear. What is interesting is that unique information is always embedded or encoded within the way the oscillating sine wave is fractalized.

just as it is encoded
in the fractalized, dimensional lines
of a mountain

Radio waves carry information in much the same way. A pure oscillating sine wave of a particular frequency is created and that wave is disturbed, the smoothness of its line fractured, by the particular kinds of information that the radio station puts into it. (Waves on the ocean are a visual example of oscillatory fractalized sine waves. They move up and down—oscillate—and their surfaces are rough—fractalized.) Radio receivers, when tuned to the frequency of the original wave, are able to decode the sounds (and information) embedded in the disturbed waves they receive, and we get to hear the weather report thirty miles from the station.

Every time life flows through something, whether matter or a part of the electromagnetic spectrum, it fractures it, turns it into a fractal. But the *way* in which life fractalizes whatever it flows through embeds unique information. (And these fractal lines are always in flux, no matter how solid a mountain appears.)

Frequency is plural unity. Frequency is a multicyclic fractionation of unity.

— BUCKMINSTER FULLER

Living systems, just like radio receivers, are supremely able to receive—and decipher—electromagnetic waves. But unlike radio receivers, they are always working at broadband, not narrowband, frequencies. Broadband means everything in the electromagnetic spectrum, not just the narrow range of electromagnetic signals we humans have tended to use for our televisions and radios. "Electromagnetism," Joseph Chilton Pearce comments, "is a term covering the entire gamut of most energy known today, from power waves, that may give rise to atomic-molecular action, to radio waves; microwaves, and infrared, ultraviolet, and visible light waves; from x-rays to gamma rays."[1]

Living organisms are extremely sensitive to all the different electromagnetic phenomena that exist, and they are able to decode the information embedded in every kind of fractalized wave they encounter. And everything that is has an electromagnetic dimension to its nature.

Each chemical element is uniquely identifiable in the electromagnetic spectrum by its own unique set of separately unique frequencies.
— BUCKMINSTER FULLER

Some of the information encoded in fractalized electromagnetic waves has nothing to do with the living organisms that encounter it, so they ignore it, just as we ignore background conversation at a party. But when it has something to do with them, it gets their attention. (Just as our attention is captured when we encounter fractalized sound waves at a party, when we hear someone mention our name across a crowded room.) Living organisms amplify these meaningful electromagnetic waves and decode them in order to hear them (just as we do). They then use the information and respond to the sender through their own uniquely fractalized electromagnetic communications. For all living organisms are transmitters as well as receivers; these communications always go both ways.

Electromagnetic waves always close back upon themselves. Deliberately nonstraight lines are round-trip circuits.
— BUCKMINSTER FULLER

CELLS AND ELECTROMAGNETIC WAVES

When a cell forms, one of its major parts is its exterior or plasma membrane. This plasma membrane is a primary sensory organ for all cells. It possesses thousands of receptors across its surface, designed to detect perturbations, influxes of chemical, electric, magnetic, hormonal, pressure, and mechanical impulses, among other things. (In a way, a cell looks sort of like a floating mine, a sphere covered with sensory bumps that respond when touched.) The cell membrane mediates the responses of the cell to all of these influxes—including those that are electrical.

One of a cell's major responses to certain electromagnetic activity is to open or close tiny ports or doorways in the membrane surface. This allows things into and out of the cell. All of these tiny ports are electrically activated. Each cell, in fact, contains thousands to millions of these ports, called voltage-gated ion channels, and they are categorized by the type of electrolyte or ion that they allow to pass into and out of the cell. The opening and closing of these ports are triggered when the cell detects and decodes the electric field of electrolytes, such as ions of calcium (Ca+), potassium (K+), and sodium (Na+). This cellular ability is very sophisticated: Cells can recognize extremely subtle differences in electric fields, noting their waveforms, amplitudes, and frequencies. They then decode them, decide how to respond, and initiate a response.

Cells also recognize other communications, other languages from the environment, not just the electromagnetic. These include such things as pressure and magnetic, chemical, and temperature fluctuations. And this sensitivity to subtle communications in the electromagnetic spectrum is not limited to cells. Even enzymes and molecules can recognize and process different electromagnetic frequencies and amplitudes. These kinds of oscillations, or wave signals, make up one of the primary languages used by all self-organized systems.

It is all a matter of vibrations—a matter of response to vibrations. In no other way than through vibrations do we get anything. You know the camera plate is struck with blows of light to burn into the sensitized surface of the picture you want to take. If you make what is called a time exposure the blows are gentle, but sooner or later they make a dent in the gelatine. The lighter parts are burned deeply, and the shadows and black places are only just touched. But it is the steady tap, tap, tap of the rays of light that do the work. We are all made— plants and fish and cats and elephants and men—of organisms built of tissue that is built of cells. The life force is in the cells—protoplasm, made up of almost everything in the universe in infinitely minute particles. Now, because that protoplasm . . . is made up of almost everything in Nature, it responds to almost everything in Nature. Protoplasm is the sensitized film on our bodily and intellectual plates; vibrations from about us strike it and gradually they make a dent.
— LUTHER BURBANK

The constant opening and closing of these voltage-gated ion channels (which occur billions upon billions of times each second of the day), the movement and activity of electrically charged ions at the surface of the cell, and their passage into and out of cells, generates constant electric fluctuations inside and across the surface of all cellular organisms. All self-organized systems have an electromagnetic identity, a field of force surrounds them, which comes from this constant electromagnetic activity. And these fields of force are always coming into contact with each other. Because every encountered field of force contains so much information about the impacts it might have on a self-organized identity, all organisms have a well-developed capacity to detect, transduce, process, and store the information contained in electromagnetic signals.

In fact, organisms at all levels of complexity generate and use electric fields in their development, their functioning, and their responses to external perturbations. These electric fields are not only active in the closing and opening of cells, but also, for example, in tissue organization. Electric fields produced by embryos are used to actually *direct* the placement and differentiation of the various cells that will become organs, the skeletal system, and so on. Millions of other functions of living organisms are also based on electrical signals. Healing is one of them.

When skin is abraded, the normal electrical potential between inner and outer skin layers is short-circuited. And "the wound," Paul Gailey notes, "provides a low-resistance return path, and the resulting electric field directs the migration of keratinocytes (new skin cells) toward the injured area. . . it is a stunning example of a self-directed organization that is globally mediated by an endogenous electric field."[2]

To facilitate all this, cellular membranes have extraordinary electrical properties. When a cell exposed to an electric field on one side of the cell depolarizes, the other side hyperpolarizes, creating what is called a dipole—a system with positive and negative fields on its opposite sides, much like an electric battery. This creates a tiny charge across the cellular membrane (between the inside and outside of the cell), giving the cell electrical potential, that is, energy it can use for carrying out its functions. (Many people now call these charges "action potentials.")

While these electrical signals are used by all organisms to promote health and functioning, they are also exchanged between organisms throughout the living world.

time for a metaphor

Imagine a woman taking a trip in a car to visit her daughter. While she is driving she gets bored and decides to listen to the radio. She turns on her favorite radio station and one of her favorite songs is playing. The signal is strong and clear, and she begins to hum to herself as she drives. But as the miles go by, the radio transmitter gets farther away from the car, and slowly, the radio signal starts to weaken. There begins to be a bit of static in the music. And, of course, the more distance that is traveled, the worse this gets. But the woman really likes this particular radio station and the songs it plays, and she is not willing to give it up. She is not going to change the channel. So she keeps listening and, as she drives, the sound keeps getting worse. She can hear less and less of the music that she loves.

Part of the problem is that the radio itself is interfering with the signal from the radio station. As the radio signal grows weaker, the electromagnetic emissions that the radio components produce begin to get stronger in relation to the signal. The signal-to-noise ratio (SNR) is approaching 1; that is, the radio signal and the electronic noise in the radio are becoming equal in strength. The higher the SNR, the stronger the signal.

Radio waves are much like waves on the ocean; they have crests and troughs, they are oscillations. But instead of water, they are oscillations of electromagnetic energy. And the radio station itself ("1500 on your radio dial") is an oscillation at a particular frequency in the electromagnetic

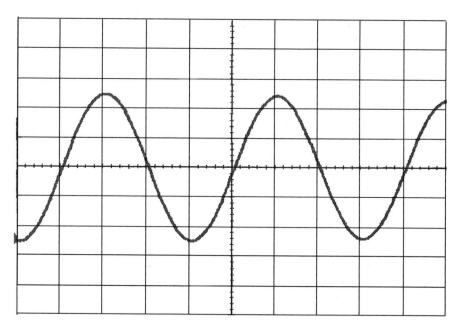

Figure 3.1. Your average oscillating sine wave

spectrum. Oscillating waves, whether of water or electromagnetic pulses, look like the picture in figure 3.1. And as with all oscillations, each peak of this kind of wave is higher than the trough. In fact, if you draw a line lengthwise through the middle of the oscillating wave in figure 3.1 the distance from the midline to the peak and the midline to the bottom of the trough will be equal. Each peak is exactly the same height as the trough is low.

(Actually, if you turn the illustration upside down, you will see that each peak is a trough and each trough a peak. The names we give them are linear expressions of something that is not linear. There really is no up or down with radio waves and, of course, they are not lines at all. They do not flow through two-dimensional planes but through multi-dimensional space, in all directions at once. And all this is only a metaphor anyway; it isn't real.)

> *I have been thinking of the difference between water*
> *and the waves on it. Rising*
> *water's still water, falling back,*
> *it is water, will you give me a hint*
> *how to tell them apart?*
> *Because someone has made up the word "wave,"*
> *do I have to distinguish it from water?*
>
> — KABIR

As the car's distance from the radio transmitter grows, the peaks and troughs of the radio signal get lower and lower; their *amplitude* drops. Amplitude is another way of saying how high and low the peaks and troughs go. What amps (or amplifiers) on stereos and electric guitars do is make these peaks and troughs very large so that their signal is extremely strong (and loud).

As the SNR approaches 1, the peaks and troughs begin to decrease more and more, getting closer and closer to the random electric (background) noise of the radio itself. (see figure 3.2.) The information (the music) contained in the peaks and troughs begins to be lost. Every time a peak or trough decreases enough to get underneath, or behind the level of the background noise, you get static. Because the peaks and troughs of electromagnetic waves, whether signals or background noise, fluctuate, aren't mathematically regular or uniform in size, some of the peaks and

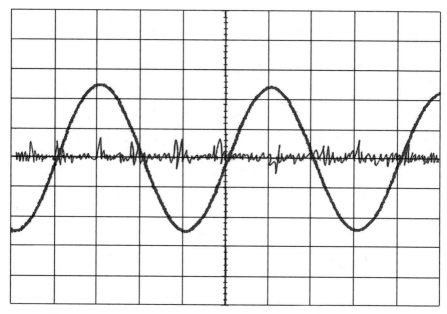

Figure 3.2. Oscillating frequency wave with background noise at midline

troughs continue to emerge above the background noise and you can still hear some of the music, though now with a lot of static. Eventually the background noise will be so loud that the oscillating waves will be unable to stick up beyond it at all, and the signal will be completely lost. When the SNR is high, the frequency wave of the signal stands out top and bottom from the background noise (figure 3.2); when it approaches 1, the oscillatory wave of the signal begins to be immersed in the noise. And though the signal is still there, it is hidden under the noise of the system itself.

But this woman happens to have a container of liquid helium in the car. She knows that if electronic components are made very cold they will make less noise. So she immerses the radio in liquid helium. As the radio components chill out, they make less and less noise and the SNR once again increases. Because the SNR is high again, the radio reception improves, and she drives on, once again happy. But as she continues to drive, she gets farther from the radio signal again and, slowly, the SNR again begins to fall. Now the problem is the noise in the antenna, it is so loud compared with the signal that only a part of the signal is getting through.

But the woman also knows that if she puts more antennas on the car, the signal will get better. And she does happen to have a bunch of antennas in the back seat.

This gets less real by the minute, doesn't it?

So she stops the car and puts antennas all over the roof. But before she reconnects them to the radio she does something else; she hooks them up in such a way that the incoming signal from each individual antenna is *averaged* with the signals from all the others before it gets to the radio.

Because of the different locations of the antennas on the roof, each antenna is now picking up more or less of the signal than every other one. The car is so far away from the transmitter now that no one antenna can get a complete signal by itself. Each piece of the signal that each antenna picks up is combined with those of the others, kind of like a jigsaw puzzle, and the composite picture that emerges is more comprehensive, fuller, than the signal any one antenna can get by itself. And it is this combined signal that is routed to the radio. The more antennas the woman puts on the car, the better the signal will be.

So the signal improves again, and the woman drives on. Still, her daughter lives very far away, and as she continues her travels, the signal eventually begins to weaken again. The radio signal is now so weak that it is beginning to be lost in the background noise from other radio transmitters in other locations that are broadcasting at the same frequency, as well as the general electrical noise from the environment. The SNR to all this background noise approaches 1 and the oscillatory wave of the radio signal begins to drop down under it, to be "drowned" out. Alas for the woman, there is nothing more that she can do except change the station.

Living organisms are, however, a lot more complex than people and cars and radios.

Many living organisms use similar processes to enhance electromagnetic signals from the world. Fish, for example, can detect extremely weak electric signals in order to hunt the fish that they eat as food.

On the exterior surfaces of their bodies, large groupings of cells are connected together into a *signal array,* much like the antenna array on the woman's car. Large numbers of cells—billions of them—hook together, combining thousands to millions of individual ion channels in order to facilitate the detection of weak electrical signals. For if one cell can detect the very weak electrical charge on a potassium ion, millions—or even billions—of them, joined together into an array, can detect electrical fields with much greater accuracy, even fields that are extremely far away.

These cells are connected together through what is called a *gap junc-*

tion. A gap junction is a tiny pore that provides a direct ionic pathway from the cytoplasm of one cell to another. Heart cells are one prime example of this phenomenon. The heart cells form gap junctions when they come into contact with one another, and they connect so well that the grouping of cells behaves as a single giant cell with a single beating frequency. A heart is, in fact, a large self-organized grouping of cells.

Cells, operating individually at their own frequencies, can synchronize or *entrain* as they move into proximity with other cells. They, like molecules, can self-organize, forming one macroscopic, ordered whole that is more than the sum of its parts. Such cellular groups tightly couple together and form aggregates that display long-range coherence or self-organization. They also develop emergent behaviors that are unique to the whole.

Aggregates of cells joined with gap junctions between them, like heart cells, are so extensive and so closely coupled that they are supremely able to sense extremely weak electric fields at close to the theoretical limits of any system to do so. Fish use exactly this process in their bodies to produce their extremely sensitive electrical detection arrays.

The more cells an organism hooks together, the less time it takes for the organism to detect a weak electrical membrane perturbation. The more ion channels involved, the faster signal detection occurs. To make this all even more sensitive, the rate of ion gating (opening and closing of openings) in these cells plays a crucial role: The faster gating rates provide more noise averaging during any time interval, and thus a stronger signal. Fish like sharks and rays not only hook together cells to make arrays, they also have a great number of arrays, all spread out over the surface of their bodies. Thus they achieve signal averaging with individual arrays and also among many arrays. This significantly increases their ability to detect a weak signal against background electrical noise. The number of cellular arrays in fish such as sharks, rays, and paddlefish is so large that the detection of an electric field perturbation takes place in about one millisecond (one-thousandth of a second)—very fast indeed. In addition, these fish can alter their internal temperature (through both natural fluctuations and interaction with the surrounding water temperature) to reduce the amount of internal electrical noise that their normal physiological functioning produces.

To gain an idea of just how sensitive to weak electrical signals these kinds of fish are, if you connected wires to each end of a $1\frac{1}{2}$ volt flashlight battery and placed the other ends of the wires two thousand miles apart in the ocean, sharks and rays would be able to detect the electric field that is produced. They can, in fact, perceive a change in an electrical field equivalent to one-millionth of a volt. Some fish have been found to be sensitive to fields as tiny as 25 billionths of a volt. This sensitivity is nearly refined enough for the fish to count individual electrons as they touch the surface of its skin.

Because all living organisms give off electrical signals as a result of their physiological functioning, any fish swimming in water also gives off weak electrical signals.

and salt water is a very good conductor of electrical signals

Paddlefish (and sharks and rays) can not only detect the weak signals themselves, but can also tell from them just what kind of fish they are sensing and whether or not it is their preferred food. They can tell how many fish there are, their size, age, and level of health; they can also pinpoint the location of the fish so accurately that they can find them in the extremely large ocean in which they are swimming.

It was long thought

by scientists

that living organisms could not detect such extremely weak fields. This is because there are so many different electric fields in the world, and they produce a lot of noise: All living organisms in the world, and there are trillions of them, are giving off electric energy; the billions of coupled cells within any organism trying to detect weak electrical signals also give off a lot of electrical energy; as water moves across the magnetic field of the Earth it causes a slight electric current; then there are electric storms, and so on and on.

To an electrical detection system, all of this background electric energy is "noise,"—electrical emissions that are not related to the signal they would like to detect. For paddlefish, sharks, and rays, it is noise that is *not* a fish they would like to eat. To make this even more complicated, biological tissues strongly shield electric fields. A living organism, such as a person, shields itself so well from electricity that a 1,000 V/m (volts per meter) external electric field will produce an electric field of only about .001 V/m inside the human body, a reduction of six orders of magnitude.

The combination of this shielding and all the background electrical noise seemed to reductionistic thinking to be an insurmountable barrier to an organism's ability to detect weak electric fields. But living organisms can and do extract information from extremely weak incoming electrical signals against this background of natural noisy electrical processes. They are able to take an electrical pulse, just as our radios and televisions do, and turn it into usable information, just as our radios produce sounds and our televisions pictures.

To detect these extremely weak signals, fish do not depend solely on temperature fluctuations and signal averaging. There are also numerous, tightly coupled cellular groupings in their bodies that respond to detected signals by oscillating with them.

> *Behind every cause lies countless other causes. Any attempt to trace these back to their sources only leads one further away from an understanding of the true cause. . . . Nature has neither beginning nor end, before nor after, cause nor effect. Causality does not exist. When there is no front or back, no beginning or end, but only what resembles a circle or sphere, one could say that there is a unity of cause and effect, but one could just as well claim that cause and effect do not exist.*
>
> — MASANOBU FUKUOKA

This has the effect of increasing the amplitude or height of the electric signal's wave, and thus of *amplifying*, or making stronger, the signal. But an even more elegant amplification process occurs through what is called stochastic resonance, or SR. *Stochastic* means "noise."

STOCHASTIC RESONANCE

When living organisms couple together a tight array of synchronized, oscillating cells, these cells are able to use the background noise itself to increase the amplitude of a weak external signal they are interested in perceiving.

With all weak signals and background noise, there is a boundary line where two events occur: noise approaches the level at which it can override any incoming signal, and the coupling strength of the cellular array is too weak to apprehend any incoming signal. At these kinds of thresholds, living organisms are extremely sensitive to perturbations or informational pulses. Because the information that they can detect in such pulses may have a strong bearing on their ability to maintain their dynamic equilibrium and thus their self-organized state of being, it is essential that living organisms be able to detect perturbations at the lowest level possible. And they have developed ways to do so, at the mathematically calculable limits of such detection. One such method is stochastic resonance.

If we go to Nature and inquire into her processes we discern more than one glimmer of light. The truth is that life is not material and that the life-stream is not a substance. Life is a force—electrical, magnetic, a quality, not a quantity.

— LUTHER BURBANK

Life has never been limited to the narrowband emissions that we use for radios and televisions; it has always used broadband—the whole electromagnetic spectrum. The number of frequencies that carry information that life can detect is thus exceptionally large. All background noise occurs in broadband as well.

Some of this background noise will naturally oscillate at the same frequency as the weak signal the organism wants to detect. What happens during stochastic or noise resonance is that these similar background frequencies are merged with the weak signal within the perceiving organism itself in order to increase the amplitude or strength of the weak signal. The noise and the signal begin resonating together. In essence, they lock together and spontaneously oscillate in harmony, much like molecules do, and form one coordinated, synchronized whole. As soon as this occurs, a powerful feedback and "feedforward" process begins, which enhances the signal. All organisms, in fact, possess mechanisms to "fine tune" their internal electromagnetic dial in order to enhance the signals they receive. As the signal is fine tuned, it gets stronger, and more and more background noise entrains to the signal, making it stronger still. This process is so powerful that a weak signal's strength can be increased ten thousand times.

We can grasp immediate causes and thus find them easiest to understand; this is why we like to think mechanistically about things which really are of a higher order. . . thus, mechanistic modes of explanation become the order of the day when we ignore problems which can only be explained dynamistically.

— GOETHE

The ability of living organisms to utilize background noise to enhance signal reception is inherent in these self-organized and emergent systems' tremendous sensitivity to external perturbations. And because all living organisms have developed in a sea of electromagnetic signals, because in many respects all things really are *only* discrete frequency oscillations, all organisms are intimately acquainted with these signals. They have learned to use them automatically, just as our lungs automatically separate out oxygen from the atmosphere and our bodies use it for functioning. All living organisms are evolutionarily accustomed to background noise, all were expressed into form within such an environment, and so take advantage of it to facilitate the detection and decoding of weak signals.

All physical phenomena, from the largest to the smallest, are describable as frequencies of discrete angular occurrence of intimately contiguous but physically discontiguous events.

— BUCKMINSTER FULLER

MAGNETIC FIELDS

Cells and living organisms not only perceive, decode, and respond to extremely weak electrical signals, they perceive, decode, and respond to magnetic signals. And magnetic fields contain information, just as electric signals do.

All living organisms create and give off magnetic fields, just as they do electric fields. Such fields strongly affect living organisms and their functioning, for living organisms and all their parts, even a single enzyme molecule, are able to detect magnetic fields and the information they carry.

Molecules of cell membranes, over millions of years of evolution, have acquired the ability to sense, decipher, and respond to low-level magnetic fields, in the form of either periodic or randomly fluctuating signals.

— TIAN TSONG

Magnetic fields have been found to directly influence a wide variety of physiological processes: enzyme activity, biological signaling, cell growth and metabolism, and tissue repair, among others. These small, field-induced changes at tiny microscopic levels have profound impacts. They cascade upward, translating into biological changes observable at the macroscopic level—in other words, at the larger level of the whole organism.

And these tiny magnetic signals, just like electric signals, can be amplified. Biological cells not only amplify the magnetic fields they encounter, but also rectify the signals they pick up, increasing their coherence. This capacity to interpret and respond to information encoded within magnetic fields is built into all biological systems and is an evolutionarily intended ability.

It has long been known that salmon, pigeons, and honeybees can sense the geomagnetic field lines of the Earth in order to orient themselves in their environment. Many birds use the Earth's magnetic lines to guide themselves to their proper destinations during migration. This sensitivity of birds, fish, and bees to magnetic fields is remarkable given the relative weakness of the Earth's magnetic field.

Magnetic fields are measured in units of tesla. The Earth's magnetic field is only about 50 microtesla (50 millionths of a tesla). In comparison, the magnetic field from a tiny toy magnet is about 1000 times greater than the magnetic field of the Earth, about 50 millitesla (50 thousandths of a tesla).

Closer examination of magnetic-sensitive birds, bees, and fish has revealed that they all contain magnetite in their bodies. Magnetite is an ore that is very sensitive to magnetic fields. (Lodestone is a form of magnetite used to make the earliest compass needles. Lodestone, unlike magnetite, is polarized. One side of it is pulled toward the magnetic north pole.) It turns out that the presence of magnetite in living organisms is pervasive, from bacteria to mammals. Magnetite is, in fact, made by living organisms inside their bodies, not gathered in from the environment. It is under precise biological control. Though most people do not know it, human beings have magnetite in their bodies, as well. It is located in the hippocampus and that organ is very sensitive to fluctuations in magnetic fields.

Weak magnetic fields modulate the rhythmic oscillations of the hippocampus. That is, they change its functioning. When the hippocampus detects a magnetic field, it decodes the information within it and shifts its functioning in response. It is more responsive to extremely low magnetic frequencies in exactly the range of the Earth's magnetic field than it is to high intensity fields. In fact, hippocampal tissue is capable of discriminating among different magnetic frequencies, between, for example, 1Hz and 60Hz oscillating magnetic fields. (Sixty Hz is the frequency at which most man-made electricity oscillates.)

The hippocampus is a very important organ for human beings. It is highly involved in interpreting spatial relationships, memory, and the extraction of meaning from the vast sea of signals within which we live. It is also closely attuned to the healthy functioning of the heart.

Physiologically, the hippocampus is a primary target for the molecules that carry information, such as ion balance, blood pressure, immunity, pain, reproductive status, and stress. It is directly involved in the feedback system for blood pressure, the hypothalamus-adrenal-pituitary axis, and the immune system. The hippocampus also works closely with the amygdala, another part of the brain, to modulate body physiology in response to emotions.

While it was long thought that the brain created no new neuronal cells after birth, it is now known that the body constantly sends stem cells to the hippocampus to be made into new neuronal cells. In response to some emotions, such as anger and fear, the body produces a great deal of cortisol. The more cortisol, or sustained negative stress, that occurs the more the ability of the hippocampus to do its job decreases. Nerve cell generation in the hippocampus slows or stops with sustained cortisol levels.

But the most interesting thing about the hippocampus is how it works with meaning. All the sensory systems of our bodies converge in the hippocampus; all the sensory impulses we receive flow to it. And all these sensory impulses contain a great deal of information. The hippocampus deciphers the meanings in the sensory impulses we receive and acts as a central transfer point for many different sites in the neocortex that, together, represent or hold memories.

In other words, the hippocampus extracts patterns encoded within sensory flows and sends these decoded patterns to other portions of the brain for storage as memories and for further processing.

Where bees and pigeons use the magnetic field fluxes to decode the information they need to orient themselves in space and direction, people use the hippocampus to orient themselves within the flow of meaning in which they find themselves daily. Once the meaning is determined, it is encoded as memory. The stronger the emotional flow that accompanies the meanings, the more strongly they are encoded as memory.

The meanings encoded in language (or in any communication, such as facial expression) cannot be decoded if the hippocampus is malfunctioning. The hippocampus decodes and integrates sensory information to provide not a directional map, but a map of experience, a map of the meanings through which we travel. The hippocampus is not only able to sense a body's orientation in space, it senses the human's orientation within *meaning*. And, in a sense, this is what salmon, honeybees, and birds do as well. They orient themselves within *directional* meaning.

The hippocampus is most active when the sensory data it receives comes from the real environment. It is designed to work—not surprisingly—with complex, nonlinear environmental information, as opposed to linear information like mathematics or what comes from a television.

> *Each factor is meaningful in the tangled web of interrelationships, but ceases to have any meaning when isolated from the whole. In spite of this, individual factors are extracted and studied in isolation all the time. Which is to say that research attempts to find meaning in something from which it has wrested all meaning.*
>
> — MASANOBU FUKUOKA

New hippocampal neurons form in response to demands on the hippocampus to process complex, nonlinear information received from the environment. The greatest degree of change or plasticity anywhere in the brain is, in fact, in the hippocampus. Enriched environments stimulate the formation of many more neurons than simple environments do. We are made to be in the wild nonlinearity of the world, and this immersion is needed for the hippocampus and our central nervous system to be healthy.

In short, what this means is that all biological systems, including human beings, are highly sensitive to both electrical and magnetic fields. The vast majority of the electrical and magnetic signals given off by living organisms—including the Earth—contains information. Whole organisms, and not just their parts, give off electrical and magnetic signals throughout their lives. These fields encode highly sophisticated information about the organisms; all living organisms have been embedded within these types of fields for their entire evolutionary history, for all the billions of years that life has existed on Earth.

And living organisms have learned to do more than simply use these fluctuating fields as part of their physiological functioning or for tracking prey. They also use them to communicate with each other. They pick up electrical and magnetic field communications from one another, alter their functioning in response, and send back responses encoded in the fields they themselves give off. In response, the other organisms alter their functioning and respond in turn. There is an extremely sophisticated electric and magnetic communication that is going on all the time among trillions and trillions of organisms. A web of communications that is so complex and detailed that there is no way to understand it with the linear, analytical mind.

An organic being is so multifaceted in its exterior, so varied and inex-haustible in its interior, that we cannot find enough points of view nor develop in ourselves enough organs of perception to avoid killing it when we analyze it.

— GOETHE

We, as human beings, are also of this Earth and possess, as all living organisms do, the ability, however atrophied, to understand these communications and respond in turn. What so many New Age practitioners call the "energy" of a thing turns out to be, in fact, the energy of a thing. It is the electric and magnetic signaling that all living organisms give off, not only as part of their physiological functioning, but as part of a complex signaling network among all life forms on Earth.

There may be no absolute division of energetic Universe into isolated or noncommunicable parts.

— BUCKMINSTER FULLER

While machines to pick up, decode, and respond to these signals may eventually be developed, human beings have always possessed one of the most powerful instruments ever created to do this—the human heart. For the human heart is vastly more than a muscular pump—it is one of the most powerful electromagnetic generators and receivers known. It is, in fact, a highly evolved organ of perception and communication.

THE HEART

The darkness of night is coming along fast, and
  the shadows of love close in the body and the mind.
Open the window to the west, and disappear into the
  air inside you.

Near your breastbone there is an open flower,
Drink the honey that is all around that flower.
Waves are coming in;
there is so much magnificence near the ocean!
Listen: Sound of big seashells! Sound of bells!

Kabir says: Friend, listen, this is what I have to say:
The Guest I love is inside me!

— KABIR

CHAPTER FOUR

THE PHYSICAL HEART

THE HEART AS AN ORGAN
OF THE BODY

The creature of institutions, bigoted and a conservatist, can say nothing hearty. He cannot meet life with life, but only with words.

— HENRY DAVID THOREAU

The transfiguration of our Western culture into an industrial egalitarianism with materialistic values first required Harvey's transformation of the heart. The [spiritual heart] had first to become a machine, and the machine become a spare part, interchangeable from any chest to any other.

— JAMES HILLMAN

One does not need a camera or tape recorder to approach birds in the field. No amount of research will help one to approach closer to them. No matter how much one investigates the bird's heart, the effort will be wasted. But by dispensing with such investigations, one will begin to understand the feelings of birds.

— MASANOBU FUKUOKA

THE MAJORITY OF MODERN PEOPLES, if asked to find the place within their body where the unique self resides, would say that they live about an inch above their eyebrows and about two inches into the skull. But most indigenous and historical peoples would locate the self someplace very different. They would gesture in the region of the heart. For most of our history of habitation on Earth, that is where the seat of intelligence, the seat of the soul, was located. That this has changed is more an expression of how and what we are taught in Western cultures than of some exact truth. For consciousness is highly mobile and is able to use a variety of locations in the body through which to process the information we receive from the world. The location that most people now identify as themselves, oriented in the brain, is only one of them.

Interestingly, as human consciousness orients itself in different locations in the body, its *mode of cognition* changes as well. The verbal/intellectual/analytic mode of cognition so common among scientists is the mode utilized by the brain. It is linear in nature. We have been so habituated to this mode that we often forget there is another form of cognition. This second mode is the holistic/intuitive/depth mode of cognition. And when the seat of consciousness is located not in the brain, but in the heart, it is this second, more holistic mode that is activated. While most people have a feeling for what this means, to gain a deeper understanding, it is important to really look at what the heart is, what it does, and just how sophisticated it can be. For the heart simultaneously operates on multiple levels of functioning.

At its most basic, the heart is a pump, circulating blood and generating pressure waves throughout the body. But the heart, it turns out, is much more than a muscular pump (and there is some question as to whether it is a pump at all). It is an electromagnetic generator, producing a wide range of electromagnetic frequencies; an endocrine gland, making and releasing numerous hormones; and a part of the central nervous system. It is, in fact, a brain in its own right.

The heart processes and generates complex patterns of multiple physiological events: it sends hormonal, neurohormonal, electric, magnetic, and chemical messages, as well as information about temperature and pressure, to the brain and throughout the body. All of these have deep impacts not only on our physiological functioning and health, but also on how and how well we think and feel—in fact, on our consciousness.

THE PUMPING HEART

The heart beats one hundred thousand times a day, 40 million times a year, some 3 billion times in the seventy to eighty years of a human life. Two gallons of blood per minute, one hundred gallons an hour, travel through vessels and arteries with a combined length of sixty thousand miles (more than twice the circumference of the Earth). So, at the most basic level, the heart is a tremendously powerful and long-lived muscular pump. Normally located just to the left of the center of the chest, it is really two pumps in one. These two pumps are the left and right sides of the heart. They sit side by side and are separated by the septum, a wall of thin tissue.

> *The dead heart was born into Western consciousness, according to Romanyshyn, at that moment when Harvey conceived the heart to be divided.*
>
> — JAMES HILLMAN

Each of these two sides of the heart have an upper collecting chamber (called the atrium) and a lower chamber (the ventricle) into which blood is received and from which it is expelled. The right atrium collects oxygen-depleted blood from the body; the right ventricle sends it to the lungs for oxygenation. The right heart's pumping mechanism is so strong that, if it were hooked to a hose, water would shoot a foot into the air. The left side of the heart is even more muscularly powerful. It collects the oxygen-charged blood from the lungs and sends it through the sixty thousand miles of blood vessels under enough pressure to shoot water six feet into the air. When the oxygen in this blood is depleted, it circulates back again to the right atrium, which then sends it from the right ventricle to the lungs for reoxygenation.

The blood pressure that physicians measure is usually given as two numbers, one after the other, for example: 120/80. The first refers to the pressure generated by the left side of the heart as its ventricle contracts and increases the movement of blood through the body. The second refers to the standing pressure in the system when the left ventricle relaxes and blood refills the collecting chambers of the heart. They are called, respectively, *systolic* and *diastolic*.

However, blood pressure is a result not only of the force of the heart's contractions, but also of the *resistance* in the vascular system to the pressure exerted by those contractions. For example, how tightly the blood vessels are constricting determines how much pressure is in the system. So, blood pressure is created by the tension between the pressure of the heart's constriction and the total peripheral resistance to that constriction. Blood pressure, on a strictly mechanical level, changes when there is a fluctuation in either cardiac output or peripheral resistance, or both. The level of pressure in the cardiovascular system is detected by the body's sensitive pressure detectors—called mechano— or baro receptors—that are scattered up and down the arterial tree.

When the valve opens between the heart and the aorta (the large blood vessel that carries blood away from the heart), the contraction of the left ventricle pushes blood into the aorta. This produces an immediate, strong flow of blood that presses against the aortic walls, causing a rapid swelling or distention of the vessel walls. There are an extremely high number of baroreceptors in this region of the cardiovascular system. With each heartbeat, the baroreceptors receive a burst of pressure impulses. These take in the information encoded in the timing, force, volume, and pressure of each successive pressure wave. They then send the resulting signals along nerve pathways to the brainstem and the central nervous system.

Each contraction of the left ventricle is, however, slightly different. The heart subtly alters these contractions in response to both external and internal information. Even though the shifts in contraction (and subsequent pressure-wave formation) are extremely tiny, the brain and other organs in the body have the capacity to take in and process the alterations for the information encoded within them. They alter their functioning in response.

But the peripheral resistance in the system changes from moment to moment, just as the heart's pumping pressure does. These peripheral resistance alterations come not only from the degree of constriction in the vessels, but also from the organs that receive the blood. Organs such as the liver, spleen, kidneys, and intestines, in fact, squeeze back, or constrict (just as the vessels do), in response to the pressure waves generated by the heart. This creates an alteration in systolic pressure. The organs and vessels also generate pressure waves of their own by altering the amount of pressure they place on the blood as it passes through. These pressure waves travel back to the heart via the blood returning to the right atrium of the heart. This process creates, in fact, a reverse pressure wave that travels back from the organs to the heart. This alters the degree of diastolic or relaxation pressure in the system.

These reverse pressure waves, like those generated by the heart itself, are rhythmic oscillations that vary from moment to moment. Like the heart, the organs and vessels are constantly analyzing information from their environment and altering their functioning, and informational communications, in response. They signal back to the heart by increasing and relaxing resistance in the system. Blood pressure, as a result, shifts from moment to moment. It is a constantly changing identity, a measure of an ongoing pressure dialogue between the heart and the rest of the system.

The baroreceptors note all these minute alterations in pressure and send them to the brain, which then alters its activity. These alterations are also fed back to the heart, which then adjusts *its* functioning, changing the beating, timing, and strength of heart contractions. It is an extremely elegant feedback loop—a living dialogue—that is used to modify the mechanical beating of the heart from millisecond to millisecond.

Thus the heart's tempo changes from minute to minute and hour to hour; the beating of a healthy heart is *never* regular and predictable. Alterations in beating patterns are most pronounced when people are young and healthy; their hearts always are highly irregular and unpredictable, always open to change. While the *average* heart beats about 60 times per minute, the beating of a healthy heart may vary up to 20 beats per minute from minute to minute. Over the course of a day, a healthy heart rate may vary from 40 to 180 beats per minute in the resting, or unstressed, heart.

> As the phenomenologist Robert Romanyshyn has indicated in his lectures about Harvey's vision, the scientific outlook requires the kind of heart it sees. . . The approach to the heart by means of literal sense perception creates the mechanical heart that Harvey describes. . . [and this occurs] in scientific thought perhaps more than anywhere—because what is imagined by science is presented as if objectively real and independent of a subjective imagination.
>
> — JAMES HILLMAN

From the time that physicians internalized the steam-engine metaphor of heart function, they have asserted that heart rates are supposed to be steady state, regular and unvarying, unless stressed by something like exercise or fright. After a demand was placed on the heart, its rate, they asserted, would return to the steady state—a process called homeostasis. Instead, like all natural systems, the heart should more properly be considered to be in a state of homeodynamis—inherently nonlinear, in constant fluctuation.

As with all complex phenomena, the complete range of factors affecting heart rate cannot, and will never, be fully identified or measured. For all nonlinear systems, this irregularity makes them more highly adaptable and robust. This plasticity allows the heart, like all nonlinear systems, to cope with the changing demands of an unpredictable and shifting environment. The heart is actually a highly refined nonlinear system and, like all such systems, responds unpredictably to stimuli.

THE NONPUMPING HEART

The traditional view of the heart as a pump was engineered out of the nineteenth-century fascination with steam engines. It is merely a mechanical model of the heart and its function that reflects the reductionistic, linear thinking of Euclid and Newton. In that model, the steam engine (or heart) is the worker that provides motive force; the water (or blood) is simply a passive, nonliving substance that is forced around the system by the pump's activity. But deeper contemplation shows, and did even in the nineteenth century, that the heart, as powerful as it is, is not really the pump it is supposed to be.

The blood is not propelled by pressure, but rather moves with its own biological momentum and with its own intrinsic flow pattern.
— RALPH MARINELLI

Modern analysis of the heart has shown that in spite of the fact that the most powerful ventricle of the heart can shoot water six feet into the air, the amount of pressure actually needed to force the blood through the entire length of the body's blood vessels would have to be able to lift a one hundred pound weight one mile high. The heart is simply incapable of producing the pressure actually needed to circulate the blood. The heart is, in reality, not the pump of the circulatory system, but plays a much more subtle and elegant role, for the heart does not pump the blood. The blood, counterintuitively, moves of its own accord.

When chicken embryos are closely examined, it turns out that their blood begins to flow in a regular circulating pattern *before* the heart has developed sufficiently to begin pumping it. And the blood does not flow like water through a hose. It is not a simple stream, flowing through a tube, but instead is something much more elegant.

Blood flow through embryo vessels is instead composed of two streams. And these two streams spiral around one another in the direction of flow. Nor do these streams flow at a regular rate. Their rates, together and separately, vary, sometimes significantly. (And this difference in flow rates is one of the causes of changes in the *temperature* of the blood from moment to moment, the thermal fluctuations from varying rates of friction. The faster the flow, the more friction, and the higher the temperature.)

At the center of these spiraling streams there is . . . nothing, a vacuum. In fact, blood flow through living vessels is much more like a tornado than anything else: It is a vortex circling around a vacuum center. At all times, up to one-third of the space occupied by the blood stream is a void, a vacuum. Such a vacuum is necessary for producing a vortex.

> *The vortex in tornadoes is a very stable configuration with a vacuum center strongly held together by a centripetal force system. . . The blood has its own form, the vortex, which determines rather than conforms to the shape of the vascular lumen and circulates in the embryo with its own inherent biological momentum before the heart begins to function.*
>
> — Ralph Marinelli

The pressure that is measured as blood pressure does not come from the heart's pumping pressure, but instead is a natural result of the movement of the spiraling, vacuum-centered blood itself.

Doppler imaging of the human bloodstream has confirmed this vortex configuration of the blood, and further found that the blood, at least in the left ventricle, is composed of not two but three streams. Both the heart and arteries move spirally, actually twist, as they work, to enhance this spiral motion of the blood. The blood vessels and heart also show a series of spiral folds on their inner surfaces. This kind of spiraling in the heart and vessel tissues enhances the flow. Liquids move faster and much more easily when they travel in a spiral motion.

like water down a drain

The spirals found in blood vessels are not present in excised (dissected) blood vessels and arteries. They are only present in living tissue. The blood and vessels work together to make these spirals, whose shape changes from moment to moment in concert with the living flow of the blood.

> *Research in plant anatomy has gradually clarified the matter of the spiral vessels found throughout the plant organism. . . In our time, researchers have insisted that these vessels themselves should be recognized as alive, and described as such.*
>
> — GOETHE

Blood is composed of various parts, and these parts orient themselves differently in the vortex of the bloodstream. The heavier red blood cells orbit nearer the center of the vortex. Lighter platelets are found farther out, with a thin layer of plasma along the vessel wall. Because all the elements of the blood are separated by this centrifuge action, each spirals at a different rate of speed. Doppler echoes have shown that a wide range of frequencies are given off by the various blood particles because of their different rates of spin.

> *The Doppler effect is usually conceived of as an approximately "linear" experience. . . But the real picture of the Doppler effect is not linear; it is omnidirectional.*
>
> — BUCKMINSTER FULLER

The red blood cells themselves not only spin around the vortex, they also spin along their own axes of rotation. They spin very fast on their individual axes and bulge in response, developing more mass at the outside of the cell from the centrifugal force of their spin. They are, in fact, smaller spinning cells within a larger spinning vortex. And this shape is not constant; it alters from moment to moment, contracting and expanding in response to spin and pressure. In part, this facilitates the movement of red blood cells through the body's tiny capillaries. They, like many things in the body, are plastic, having the quality of *plasticity*.

> *The word "plastic" came a long time before the invention of plastic. It means malleable, changeable, formable.*

The heart, inserted into this flowing system, serves an auxiliary function. It couples together with the circulatory system and phase-locks its own pulsating, spiraling, pumping action with the action already occurring in the blood. This stabilizes and regulates the flow. The heart constantly monitors the blood through sensitive receptors embedded throughout the heart and arterial tree, and alters its functioning constantly to make subtle, second-to-second shifts in the flow of blood.

Just as the heart synchronizes its contractions to facilitate the movement of the blood, the blood vessels also synchronize their contractions to move the blood through the vessels and capillaries. The pressure wave from the heart's beat continues through the body, carried along by the vessels themselves as they contract in turn to help the blood flow. The skeletal muscles assist as well by contracting and expanding, squeezing the vessels even more. And, of course, the organs that receive blood squeeze too, resulting in a complex harmony of pressure waves, all of which carry information, all of which stimulate interaction. All of this is, in fact, a dialogue.

and it is all exquisitely timed

The result is that every minute, two gallons of blood flow through vessels more than sixty thousand miles long.

But the heart's presence in the circulatory system is crucial for many reasons, not just because it stabilizes blood flow and generates pressure waves. The blood, by nature of its primary function—oxygenation—has to reach every cell in the body. Thus the heart's presence in this system allows it to affect every cell and organ in the body. This facilitates its role as a primary endocrine gland and the most powerful biological electromagnetic oscillator in the body.

This heart acts not as central king or pump, but as the circulation itself, sensitive to many things in many places.

— JAMES HILLMAN

THE HEART AS ENDOCRINE GLAND

Nineteenth-century medical practitioners were excited to discover powerful glands in the body that produced substances with marked impacts on the body's functioning. These, while located at widely divergent places in the body, were grouped together in what they called the endocrine system. This included the hypothalamus and pituitary glands in the brain and the adrenal glands on the kidneys. However, it turns out that every organ in the body produces hormones, molecular substances that significantly alter physical functioning. The gastrointestinal tract produces at least seven different hormones; the pituitary gland, in contrast, produces only nine. The heart produces at least five, though more are being discovered all the time. There is, in fact, no such thing as *the endocrine system,* and contrary to most medical thinking, the heart is one of the major endocrine glands in the body.

The hormones produced by the heart have broad physiological impacts, affecting the functioning of the heart, brain, and body. The first two hormones researchers noticed were atrial naturetic peptide or factor (ANP or ANF) and brain naturetic peptide or factor (BNP or BNF). ANF is produced in the atria of the heart, BNF is produced in the ventricles. The latest discoveries are called C-type naturetic peptide (CNP), heart-produced vessel dilator (HPVD), and calcitonin gene-related peptide (CGRP). Research has recently shown that HPVD strongly inhibits pancreatic cancer cells. CGRP acts synergistically with nitric oxide to cause vasorelaxation and exert antiproliferative actions, protecting the arteries from atherosclerosis, coronary heart disease, and stroke. A number of inflammatory agents (prostaglandins, histamine, and bradykinin) and the metabolic end product lactic acid trigger the release of CGRP so that blood vessels dilate.

While BNF has a number of effects, when a person is under stress, BNF activates a specific pathway in the neuronal cells of the brain and the heart that causes the secretion of a unique protein—beta-amyloid precursor protein. This protein protects both brain and heart neurons from stressors (such as toxic levels of glutamate), especially in the hippocampus. In other words, the heart creates a specific hormonal neuroprotector to safeguard brain function, specifically targeting neuronal cells in the hippocampus.

The heart atria contain dense cellular bodies or granules similar to those in the pancreas and pituitary gland. ANF, the most studied of the heart's hormones, is stored in these granules. (The right atrium contains two to two-and-a-half as many granules as the left.)

These granules are highly concentrated near the surface of the heart and in the atria's exterior regions. When the aortic valve opens, the left ventricle contracts and blood is pushed into the aorta. The pressure and subsequent distention cause the release of ANF into the blood.

The hormone travels through the bloodstream to targets throughout the body, including vascular tissue, cerebrospinal fluid, the kidneys, adrenal glands, immune system, brain, posterior pituitary gland, pineal gland (which secretes melatonin), hypothalamus, lung, liver, ciliary body (which secrets the lymphlike aqueous humor of the eye), and small intestine. ANF also plays a role in regulating the hormonal pathways that stimulate the function and development of male and female reproductive organs.

The amount of ANF released depends on the pressure from the contraction, which is itself altered minutely with every beat, according to what the heart senses. ANF delicately adjusts the complex balance of the interconnected, self-organized system that is the human body. It alters the functioning of any organ that receives it. In the hypothalamus, ANF inhibits the release of vasopressin, a hormone stored in the posterior pituitary. Vasopressin is antidiuretic and is a major factor in the constriction of arterioles and capillaries. ANF also relaxes the smooth muscle cells of the blood vessels, causing blood pressure to lower, and inhibits the adrenal glands' secretion of aldosterone—a hormone that tends to raise blood pressure. ANF also stimulates the kidneys to increase the excretion of sodium (and inhibit its reabsorption), which also plays a role in regulating blood pressure. ANF relaxes muscle cells throughout the vascular system as well. It touches the surface of cells and activates cyclic GMP, a messenger molecule that conveys information sent by the heart through ANF release into the interior of cells. ANF helps regulate blood volume and the body's potassium levels. It binds to a number of sites in the eye, affecting ocular pressure and focus. Depending on ANF levels, eye focus may be sharply focused or relax to a more soft-focused, peripheral orientation.

this is important, as you will see

ANF induces alterations in the body's levels of a number of hormones and neurotransmitters: plasma renin, norepinephrine, aldosterone, catecholamines, cortisol, arginine vasopressin, and dopamine. ANF alters the electrolyte balance in urine, alters blood dynamics, and alters hormone production and release in the adrenal gland. It increases urine volume and changes the urinary makeup of chloride, potassium, calcium, phosphate, and magnesium, and increases albumin and free water excretion.

ANF is depleted in hypertension, and impairment of the ability of the heart to produce ANF is directly correlated to the progress of hypertension. In a sense, the dense clusters of cells in the heart that produce ANF become exhausted, in much the same way the adrenals can become exhausted. This indicates overuse of the system, an overstimulation, and can be compared to the adrenal exhaustion that occurs in type A personalities after years of high-intensity living. The heart constantly senses the makeup of the blood and alters its release of ANF to fine tune the regulation of blood volume. Perturbations in beat timing and force directly affect the timing and amount of ANF that is released.

BNF and CNP, though much less understood, both play important roles in many of the same organs. BNF and CNP are present in the cerebrospinal fluid and affect pituitary and hypothalamic function. CNP has a direct influence on adrenal function and the production of sexual hormones. But what is really interesting is that ANF, CNP, and BNF all strongly impact the hippocampus and integrated functions of the central nervous system involving

> *learning and memory, [and] exploratory activity in a new environment.*
> — GYULA TELEGDY

BNF, because it stimulates beta-amyloid precursor protein, helps protect hippocampal tissue and enhances its functions in learning and memory.

These hormones, all produced by the heart and released into the bloodstream, profoundly affect how and what we learn, how we remember, and how well we remember. The more these hormones are produced, the better we remember and the better we learn. They also play a crucial role in helping us orient ourselves in space and time, enhancing our locomotor activity. Their actions are facilitated by a number of neurotransmitters—neuronal hormones—that the heart also makes: dopamine, norepinephrine, and acetylcholine.

Dopamine is created in the heart from the dietary precursor L-dopa (or *its* precursors, tyrosine and phenylalanine), and is an essential chemical enabling the transfer of information from neuron to neuron. L-dopa is also intimately connected to sexual interest and erections in men and the ability to have orgasms for women. Low dopamine levels are at the heart of the development of Parkinson's disease.

Acetylcholine is an essential brain transmitter as well, playing a crucial role in memory. Problems with acetylcholine in the brain contribute to loss of memory in Alzheimer's patients.

Like dopamine, norepinephrine is made from the precursors L-dopa, tyrosine, and phenylalanine. Norepinephrine regulates the movement of fats in the bloodstream and the contraction of arterioles. It plays an essential role in regulating arterial health, fat processing, and atherosclerosis.

The relationship of the heart to these hormones shows how intimately involved the heart is in brain function. In many respects, it can be seen as something like a unique type of brain hooked into the central nervous system just as the brain is.

THE CENTRAL NERVOUS SYSTEM HEART

Between 60 and 65 percent of the cells in the heart are neural cells, up to 25 percent of the total cells of the heart. They are the same kind as those in the brain and they function in exactly the same way. In fact, certain crucial subcortical centers of the brain contain the same number of neurons as the heart. The heart possesses its own nervous system and, in essence, *is* a specialized brain that processes specific types of information. Heart neurons, just like those in the brain, cluster in ganglia, small neural groupings that are connected to the neural network of the body through axon-dendrites. Not only are these cells involved in the physiological functioning of the heart, but they also have *direct* connections to a number of areas of the brain, and produce an unmediated exchange of information with the brain. (*Unmediated* means that there are no interrupts in the circuit from the heart to the brain. A light switch is a circuit interrupt which can be turned on or off.)

The neural connections to the brain from the heart cannot be turned off; information is always flowing between the two. The heart is, in fact, directly wired into the central nervous system and brain, interconnected with the amygdala, thalamus, hippocampus, and cortex. These four brain centers are primarily concerned with

1. emotional memories and processing;
2. sensory experience;
3. memory, spatial relationships, and the extraction of meaning from sensory inputs from the environment; and
4. problem solving, reasoning, and learning.

(The heart makes and releases its own neurotransmitters as it needs them. By monitoring central nervous system functioning, the heart can tell just what neurotransmitters it needs and when in order to enhance its communication with the brain.)

The heart also has its own memory. People who receive transplanted hearts often take on behaviors common to the person to whom the heart originally belonged *such as liking salsa for instance*

but were not something they formerly did at all. The heart, which possesses the same kind of neurons as the brain, stores memories. These memories affect consciousness and behavior, how we perceive the world. They most often have to do with specific emotional experiences and the meanings embedded within them. The more intense the emotional experience, the more likely it will be stored by the heart as memory.

Neuronal discharge in the brain—the oscillating pattern of informational pulse release in the amygdala, hippocampus, thalamus, and sometimes the neocortex—is in phase with heart and lung cycles. These discharges are state-dependent. In other words, changes in heart activity—blood pressure, timing of beats, wave pulsations in the blood, hormone and neurotransmitter creation and release, and more—all shift the functioning of these areas of the brain. Information embedded within cardiac outputs directly reaches many of the subcortical areas of the brain involved in emotional processing. The kinds of information that the heart sends significantly shifts functioning of the amygdala

thus affecting emotions

and other subcortical centers of the brain. The kind of activity displayed in the central nucleus of the amygdala has been found to be dependent on input from the aortic depressor or carotid sinus nerves. Heart researcher Rollin McCraty comments, "Cells within the amygdaloid complex specifically responded to information from the cardiac cycle."[1]

Single neurons in the brain alter their behavior in response to the signals received from each heartbeat. In response to cardiac input, complexes of neurons in the brain change their grouping and firing patterns. They alter their behavior in order to embed the information received through cardiac function and send it into the central nervous system. The information embedded within cardiac pulses alters central nervous function in behaviorally significant ways. There is, in fact, a two-way communication between heart and brain that shifts physiological functioning and behavior in response to the information exchanged.

Analysis of information flow into the human body has shown that much of it impacts the heart first, flowing to the brain only after it has been perceived by the heart. What this means is that our *experience* of the world is routed first through our heart, which "thinks" about the experience and then sends the data to the brain for further processing. When the heart receives information back from the brain about how to respond, the heart analyzes it and decides whether or not the actions the brain wants to take will be effective. The heart routinely engages in a neural dialogue with the brain and, in essence, the two decide together what actions to take.

But even more intriguing than all of this is the electromagnetic activity of the heart. And this story starts with the unusual manner in which the beating of the heart begins. For cells in the heart, at a very early stage in the development of the embryo, spontaneously begin pulsating or beating all on their own. They suddenly synchronize, become self-organized, and display emergent behaviors.

The Electromagnetic Heart

Although most people have heard of pacemakers—mechanical contrivances that help hearts beat more regularly—the real pacemakers were invented long before men in laboratories thought of them. Natural heart pacemakers are groupings of cells that, just like groupings of molecules, display spontaneous, autonomous self-organization.

When the organizing subunits that are destined to become the heart's pacemaker cells reach a certain level of complexity during embryonic development, they self-organize. At that point, the first pacemaker cells begin beating, displaying an emergent oscillatory behavior. After the first of these heart cells begins to spontaneously beat, every new pacemaker cell that develops entrains, or synchronizes, to the first. Ultimately, there are a tremendously large number of pacemaker cells, millions of them, all working together as one beating unit, synchronized in their harmonic oscillations.

Thus, the heart can be considered as a dynamic, non-linear, harmonic oscillator.

— Rollin McCraty

As with all nonlinear systems, the coupling of millions and millions of heart cells modifies their behavior and produces new behavior and potentials that cannot be predicted from the study of the individual cells in isolation.

If one of the heart's pacemaker cells is removed from the body, kept alive, and placed on a slide,

what a horrible thing to do

it will begin to lose its regular beating pattern and start to fibrillate—to beat wildly and irregularly—until it dies. But if you take another pacemaker cell and put it close to the first—they do not need to touch—their beating patterns will synchronize; they will beat in unison. A fibrillating cell, placed next to a non-fibrillating pacemaker cell, will stop fibrillating and entrain, or begin to beat in unison with the healthy cell.

The reason that these cells do not need to touch is that they are producing an electric field as they beat, as all biological oscillators do. The mechanical motors used to generate electricity are, in fact, only a pale imitation of the powerful electrical generators that life developed billions of years before we emerged as an ecosystem expression of the Earth.

The pacemaker cells in the heart are located in a number of places throughout its tissue. (There are other pacemaker cells in other organs throughout the body that help *them* function.) The two most powerful groupings of heart pacemaker cells are in the upper right and left atria, the portions of the heart that start the contractions that force blood into the ventricles.

The electrical impulses from these cells are transmitted to the muscle tissue in the atria. All the muscle cells are connected together through gap junctions and constitute one synchronized organism. The electrical impulses flow in millisecond time intervals through the gap junctions, causing the heart muscle to contract simultaneously. This impulse signal is insulated from the ventricle portion of the heart; it can only flow into the ventricle through a special junction point: the atrial-ventricular node. There is a one-tenth second delay before the signal is transmitted so that the ventricle can completely fill with blood. Then the muscle contraction in the ventricle forces it out and into the body.

These regular muscle contractions of the heart generate volume currents—electrical pulses—in the ionic, electrically conductive tissues of the body. There is an electrical charge that is generated with every beat and conducted throughout the body's tissues. Each pulsating beat of the heart, in fact, produces two-and-a-half watts of electrical energy. And this electrical charge, though pulsating, is continuous, just as the heart, throughout life, is continuous in its beating. It is this pattern of electrical activity that is measured when an electrocardiogram (ECG) is taken by placing electrodes on the body. But the heart also generates magnetic fields (which is why it is often referred to as electromagnetic, instead of just electric) and those magnetic fields can be measured as well, with a magnetocardiogram.

When the heart valves open to allow blood to enter the atria, the blood swirls in under tremendous pressure, creating an even stronger vortex than that already present in the blood vessels. (This vortex continues throughout the heart as the blood moves from chamber to chamber.) The insertion of even a single ion into this vortex creates a powerful magnetic field. And the blood contains many more than a single ion. The heart's electric and magnetic fields are created not only by the vortex of the blood in the heart, but also by the vortex of the blood in the vessels *and* the spinning motion of the blood cells themselves as they travel through the vessels. Just as the blood carries chemicals and cells, it also carries electromagnetic signals. These, like the other blood components, travel to all parts of the body, reaching every cell.

blood is a very good conductor of electromagnetic waves

While the electric and magnetic fields of the heart are, unsurprisingly, similar in shape, the development of magnetic imaging devices over the past twenty-five years has enabled researchers to reveal that the magnetic and electric fields of the heart possess distinct differences as well. The magnetic and electric fields of the heart actually encode different kinds of information. And just as the heart's activity alters with each beat, the shape of the electric and magnetic fields of the heart changes with each beat.

The heart, in fact, produces a tremendously powerful, broadband electromagnetic field as it beats. When the electromagnetic patterns of the heart are imaged, they form patterns very similar to those given off by magnetite or a magnet. A magnet's magnetic field can be demonstrated by placing iron filings on a sheet of paper and holding the magnet underneath it. The filings will quickly move on the paper into a pattern, aligning themselves to the magnetic field that the magnet produces. But the magnetic field produced by the heart is not located on a flat piece of paper. It extends around the body in a *torus,* a fractal, sort-of-spherical shape that continually flows through space.

Measured with magnetic field meters, the electromagnetic field that the heart produces is some five thousand times more powerful than that created by the brain. While strongest at the body's surface, it extends out further than human measuring devices can detect. The most sensitive electromagnetic devices human beings know how to make can still detect it at up to ten feet from the body. (However, as with all electromagnetic waves, there is no limit to how far the heart's electromagnetic field actually travels, whether we can measure it or not.)

The electromagnetic field of the human body aligns roughly along a person's spine, from the pelvic floor to the top of the skull. This field permeates every cell in the body.

The electromagnetic field of the heart, however, is not a neatly symmetrical field of similar arcs. For this is not a linear formation, but a nonlinear one. Its shape is the expression of a constantly shifting, living process; it changes with each alteration of the heart as the heart takes in and processes information about its internal and external environments.

The heart produces a range, a spectrum, of electromagnetic frequencies. Any frequency in this spectrum can contain a significant amount of information, just as one particular frequency on the radio dial can contain huge amounts of information. And each section of the field, no matter how small, contains all the information encoded within the whole field.

The Earth's magnetic field is a very similar torus (or pattern) to what hearts (and magnets) emit. The North and South magnetic poles are the two ends of the dipole, like the lower and upper ends of our spines (or the two poles of a battery). Like that of the heart, the Earth's magnetic field is a constantly shifting, living field. All living organisms possess just such a torus, including plants. (The blood cells themselves form tiny torus-shaped fields around themselves as they spin, creating individual electromagnetic fields within the spinning vortex of the bloodstream. Thus there are electromagnetic charges nestled within electromagnetic charges.)

The whole body is cradled within the electromagnetic field generated by the heart. The information embedded within that field is communicated to the external world through electromagnetic waves reaching out from the body. It is communicated within the body through the bloodstream, which conducts electromagnetic impulses throughout the body.

Blood possesses tremendously potent electrical conductivity. Thus the blood carries more than pulse waves; it also carries electrical messages. As just one example, DNA is sensitive to electromagnetic fields. The electromagnetic fields produced by the heart are involved in the regulation of DNA, RNA, and protein synthesis—they help induce cell differentiation and morphogenesis. This is not random, of course, for the electromagnetic waves the heart produces (and the shape of its field) are much like radio waves in that the shape of the waves changes depending on what information they receive. The human body is supremely able to decode these messages, just as our car radios can decode radio station signals.

But the body's system is much more elegant than that of our radios. It is a living system. The heart sends out informational messages on multiple frequencies, all the millions of elements of the body receive them and respond—and, within milliseconds of time, the heart's beating patterns alter in response. The electromagnetic pulsations of our hearts are more properly part of an ongoing dialogue, a communication, whose function is to help maintain the dynamic equilibrium of the self-organized systems we know as ourselves.

The distinct patterns of neurological, biochemical, biophysical, and electromagnetic activity generated by minute and precise alterations in heart activity function as a language that encodes information and communicates it from the heart to the body and to the world outside the body. All of these patterns are, in fact, the mediums through which the heart maintains its dynamic equilibrium.

But the heart is not only concerned with its interior world. Its electromagnetic field allows it to touch the dynamic, electromagnetic fields created by other living organisms and to exchange energy. Like all nonlinear systems that display self-organization and emergent behaviors, the heart is supremely sensitive to external perturbations that may affect its dynamic equilibrium. The heart not only transmits field pulses of electromagnetic energy, it also receives them, like a radio in a car. And like a radio, it is able to decode the information embedded within the electromagnetic fields it senses. It is, in fact, an organ of perception.

All that a man has to say or do that can possibly concern mankind, is in some shape or other to tell the story of his love—to sing, and, if he is fortunate and keeps alive, he will be forever in love. This alone is to be alive to the extremities. It is a pity that this divine creature should ever suffer from cold feet; a greater pity that the coldness so often reaches to his heart. I look over the report of the doings of a scientific association and am surprised that there is so little life to be reported; I am put off with a parcel of dry technical terms. Anything living is easily and naturally expressed in popular language. I cannot help suspecting that the life of these learned professors has been almost as inhuman and wooden as a rain-gauge or self-registering magnetic machine. They communicate no fact which rises to the temperature of blood-heat.

— HENRY DAVID THOREAU

CHAPTER FIVE

THE EMOTIONAL HEART

THE HEART AS AN ORGAN
OF PERCEPTION AND
COMMUNICATION

The spirit of life, which hath its dwelling in the secretest chamber of the heart.

— DANTE

We evaluate everything emotionally as we perceive it. We think about it after.

— DOC CHILDRE

The intellect is powerless to express thought without the aid of the heart.

— HENRY DAVID THOREAU

Only a reductionist science would need to "prove" the ridiculously obvious: that our hearts are perceptual organs, crucial to our humanness.

— AUTHOR'S JOURNAL, NOVEMBER 2003

THE TENDENCY FOR HEART CELLS TO ENTRAIN with one another, merely because of the proximity of their electromagnetic fields, extends to any electromagnetic field that comes into contact with them. Just as the electromagnetic fields of two heart cells cause them to begin beating or oscillating in unison, when the electromagnetic fields of two hearts come together, they also begin to oscillate or entrain to each other. But this phenomenon extends even further. When the heart's electromagnetic field and any other organism's electromagnetic field

whether it has a "heart" or not

are in close proximity, the fields entrain or synchronize, and there is an extremely rapid and complex interchange of information. As the two fields harmonize with one another, shifts occur in each electromagnetic field, producing significant alterations in the physiological functioning of each organism. For not only does each electromagnetic field alter, but the information embedded within each field is also taken in by the receiving organism. The information in the encountered electromagnetic field is a perturbation of each organism's dynamic nonequilbrium and, like the clown on the unicycle, an alteration of internal dynamics is needed in order for them to keep equilibrium.

The perturbations that occur when another electromagnetic field is encountered alter each organism's coupling dynamics, producing new, cooperative, dynamic states. In addition, as the two fields come together and synchronize, the process produces a combination field, in effect, two fields in one. And these two fields are, like all nonlinear oscillators, in harmony. They produce something that is more than the sum of their parts. These fields are, as Joseph Chilton Pearce says, "aggregates or resonant groupings of information and/or intelligence."[1] A unique identity comes into being and exists as long as the two fields are synchronized.

Energy systems, like the heart, are open systems; they are always interacting with other energy systems. They are always using, storing, and emitting energy. The more complex a system is (meaning the larger the number of self-organized subunits that combine in self-organization to make it up), the more complex its energy and information processes become, and the more factors must be taken into account to maintain its dynamic equilibrium.

Within the electromagnetic spectrum, the heart must decode and encode information across multiple waves and frequencies with each beat.

At the same time, it generates and delivers different pressure waves, sound waves, thermal fluctuations, hormonal cascades, neurotransmitters, and neural bursts of information directly to the centers of the brain that it is connected to and to the rest of the body. At any one moment in time there is an informational gestalt, a gesture of communication, going out from the heart to both the external and internal environments in which it lives. And this particular gestalt changes from moment to moment, depending on the information the heart receives from those environments.

The heart is extremely complex, and the energy fields it creates, emits, and uses in communicating with other energy systems (the rest of the body or other organisms) are extremely complex as well. The pulses of energetic information that the heart sends out, for example, do not all travel at the same speed.

Like a lightning strike: you first see the flash, then hear the sound, then feel the rumble.

Some electromagnetic waves—like visible light—travel very fast. Some, like sound waves, are slower; pressure waves are slower still. All these pulses of energetic information travel at different rates within

and outside of

the body, and produce effects at different times. All of these energetic expressions encode meaning, and all have effects on external organisms. The sound of a slow, external heartbeat helps calm infants; a more rapid beating, inserted in the background score of a horror film, can generate feelings of panic in the listener.

Electrical and magnetic energy, in combination with other forms of energy, radiate from the body and travel into space as organized patterns of energy.

— LINDA RUSSEK AND GARY SCHWARTZ

The organized patterns of energy from the heart, in fact, have been shown to directly affect the functioning of organisms outside the heart.

The merging and entrainment of our hearts with other electromagnetic fields is extremely natural to us; it is one of our earliest experiences. For this entrainment first occurs before birth. We are immersed in our mothers' electromagnetic fields while in their wombs, and electroencephalogram and electromagnetogram readings have shown that the fields of the two, mother and infant *in utero,* naturally synchronize or entrain. During breast-feeding and holding, the infant's electromagnetic field is constantly resynchronized with the mother's. As Joseph Chilton Pearce remarks, "The mother's developed heart furnishes the model frequencies that the infant's heart must have for its own development in the critical first few months after birth."[2] And the mother's electromagnetic field encodes large amounts of complex information that affect the child far beyond mere mechanical dynamics. At the simplest level, how the mother feels about the child, whether the child is wanted or loved, is conveyed to the developing embryo through information encoded within alterations of the mother's electromagnetic field. Those alterations are specific embeds, encodes, of information that the receiving field of the developing child can decipher—just as a radio receiver can decode radio waves.

Because the human heart is born into a situation in which its first functionings are intimately involved with information coming from another electromagnetic field, it continues throughout its life to be sensitive to the information in electromagnetic fields. It gestates, you might say, within this kind of language. It is the heart's birth tongue. So, throughout life, the heart actively scans fields it perceives, looking for patterns of communication and information. Whenever the heart comes across other biological oscillators and their electromagnetic fields, and as its field is perturbed by the other fields at their first touch, the heart experiences an alteration in its electromagnetic spectrum. The way the electromagnetic field is altered conveys information. If the two fields synchronize, even more information is conveyed. The way these radiating fields of energy patterns and their perturbations are experienced by human beings is unique. They are experienced as emotions.

The basic colors our eyes can detect combine to make up the infinite range of colors that we can see. Each of these colors has a different waveform, a different frequency; these frequencies are taken in through the eyes, processed in the visual cortex, and interpreted as color. All our sensory mediums are similar in this way. For example, the four basic tastes—sour, sweet, bitter, and salty—combine in a multitude of ways to make up the spectrum of tastes we can experience. The electromagnetic field frequencies of the heart are experienced not as colors or tastes, but as emotions. (The slightest emotional change, due either to internal or external factors, shows up immediately as a change in heart rate and heart rate variability patterns, and vice versa.)

The heart is, in fact, an extremely sensitive sensory organ whose domain is that of feelings. Emotions represent the impact of specific electromagnetic spectrum carrier waves upon us, as colors are the impact of visual carrier waves. Like colors and tastes, the broad spectrum of complex emotions we can experience is created through subtle combinations of a few basic emotions: mad, sad, glad, and scared. These combine to form many more complex emotional states, such as jealousy, awe, and love. They combine in even more complex forms than these of course, because the number of emotions we can experience, fleeting as most of them are, cover a nearly infinite range. Just as the variations in electromagnetic response of the nonlinear oscillator we know as our heart approaches infinity through the fractalization of its processes, our experiences of those shifting processes allows a nearly infinite number of emotional blends.

INTERNAL AND EXTERNAL ELECTROMAGNETIC FIELDS

The human body contains a great many biological oscillators, all hooked together in the organism we know as ourselves.

The three most powerful are the heart,
gastrointestinal tract, and brain.

The internal energy fields we sense within us, coming from all our biological oscillators (from cells to organs to the combined, whole organism), contain certain kinds of information about our internal world. We feel that information as certain kinds or groupings of emotions. These emotions give us informational, sensory cues about what is going on within us.

if only we will pay attention

When we decipher those cues, just as when we decipher the pattern of visual cues that is a road sign, we gain information about the road we are on, the path we are taking.

That our internal world expresses information to us in emotional information pulses was reflected in classical understandings that organ malfunction would be accompanied by specific emotional states. A malfunctioning liver, for instance, was considered the source of unexplainable anger, a malfunctioning gallbladder of melancholy. Each of these malfunctions affects the makeup of the heart's electromagnetic field. Even in a healthy system, a great deal of the emotional flux we experience daily comes out of an intricate interplay between our internal subunits: molecules, cells, and organs. Studies have found, for example, that there is a relationship among splenic contraction, blood pressure, and emotional states. As their function shifts, the changing electromagnetic fields of those biological oscillators alter the heart's electromagnetic field. We then experience an electromagnetic pulse of information, felt as emotions, coming from a shift in our internal functioning. (This alteration also changes the heart's pressure waves, something traditional Chinese physicians know and have formalized in pulse diagnosis.)

Unfortunately, in our time, our languaging for these internal states is extremely limited. We may feel "under the weather," but we can feel under the weather in a great many ways, and each of these ways has a particular and unique feeling or complex of feelings attached to it. We may feel "blah" or "sick" or "depressed," but each of these statements conveys little information about our internal state. They are not elegant, specifically communicative statements. To a great extent, this limitation comes from a cultural, long-term lack of focus on the great variety of emotional states that are generated by alterations in our internal world. Ancient and indigenous cultures, focused more on the heart as an organ of perception, generally were more able to elegantly articulate these internally generated emotional states.

If we direct our consciousness outside ourselves and pay attention to the biological oscillators we encounter there, we can also become aware of the emotions generated by our encounters with external electromagnetic fields. When the fluctuating electromagnetic field of our heart touches another electromagnetic field, whether from a person, rock, or plant, we feel a range of emotional impressions that is our experience of the information encoded within those organisms' electromagnetic fields and the alterations that have occurred in our field. This is, in fact, the source of the deep feelings that come from our immersion in wild landscapes, the feelings we have when we see the Grand Canyon, for instance. And these externally generated feelings are an important and essential source of emotions for all human beings, for we emerged not only from our mothers' wombs, but also from the wildness of the world. We developed nestled not only in our mothers' electromagnetic fields, but also within the larger electromagnetic field of the Earth. We are an expression of the ecosystem, the womb, the Earth, an ecological response of the planet. And this kind of information exchange is embedded deeply within our cellular memories.

The heart is, then, a receptor organ, receiving information not only from within, but also from the external world. The heart processes the impact of external events on the organism within which it is located, changing its beating patterns, pulse waves, electrical output, hormonal functioning, and neurochemical release. These changes in function are used to impart information to the rest of the body and also to the central nervous system, the brain. The heart serves as a conductor of depth information from the external world to the central nervous system and brain, where it interacts with central nervous system functions. These cardiovascular events, or alterations, exert strong influences on cortical functioning and are specifically detectable as sensory signals. Close examination reveals that these alterations in heart function in response to external phenomena have the same kinds of effects on cortical functioning as do more classical sensory inputs, that is, visual, auditory, olfactory, tactile, and gustatory stimuli. The incoming sensory perceptions from the heart have the same ability to capture the attention and shift behavior as those five sensory mediums.

When the heart is impacted by events in the external environment, information about those external events is encoded in various cardiac wave patterns (beating patterns, pressure waves in the blood, and so on) that are analogous to the different wave forms that come from visual or auditory stimuli—light and sound waves. With visual and auditory stimuli, the cortical centers of the central nervous system take in the colors and sounds and allow the patterns of meaning embedded within them to emerge into a comprehensive whole so that they can be understood. The heart's wave forms, experienced as emotions, also have embedded meaning and this meaning can be extracted from the emotional flow just as meaning is extracted from visual and auditory flow.

Because we are trained to ignore these particular kinds of sensory cues and the information they contain, most people do not consciously utilize the heart as an organ of perception. Most of the information received is thus processed below conscious levels of cognition. Still, because the heart is such an essential organ of perception, because emotions are still crucially important to the experience of being human, so much a part of our environmental history and ecological expression, the heart's power as an organ of perception cannot be completely erased. Some people remain highly attuned to its perceptions, just as others do not. People's awareness of heart-encoded information is highly dependent on psychological and historical variables: their schooling, past relationship with their bodies, environments, and histories of emotional experiences.

Most contemporary research on external electromagnetic fields is concerned with those we encounter in other people. Here, too, our languaging is extremely limited. We use the word "love" to describe a great many different states in our heart's electromagnetic field. We may "love" broccoli, a friend, our dog, a book, getting together for lunch, or our spouse, but these various "loves" are all different. Nevertheless, our language provides few ways to easily distinguish among them. And while we may recognize that the intermingling of our own hearts with the hearts of others produces different electromagnetic states and, thus, different emotions, our sophistication with them and our ability to describe them is severely limited. The recognition that our electromagnetic fields have a natural capacity to interact and synchronize with other types of electromagnetic fields—that is, with ecosystems and members of those ecosystems—is nearly atrophied.

While scientists are excited about the knowledge they are gleaning about the heart and its functions, none of it is really new. That the heart is intimately concerned with emotions, with who we are and how we experience and are experienced by the life outside us, has been known throughout history to all the world's cultures.

Our language (like all languages) contains wisdom about the heart that we rarely call up into our conscious minds. We have all known, at one time or another, a man who is "big-hearted," a woman who is "good-hearted," and may even have friends who are "kind-hearted." If we tell them so, we may do it in a "heartfelt" way. We can eat a "hearty" meal, share a "hearty" laugh, or even look "hearty." Our profession or our mate may become the "heart" of our life, or we may work for long years to attain our "heart's desire." And because the heart does in fact act as a specialized brain, it is actually possible to "follow your heart" or to "listen to your heart."

If we are dejected or hopeless, it may be said that we have "lost heart." If a loved one rejects us, we can become "heartsick" or "broken-hearted." If we are being unkind, someone may implore us to "have a heart" or not to be "heartless." People can be "cold-hearted" and cruel or even "hard-hearted." Our hearts are intimately concerned with who and what we are, each day, and throughout our lives.

Our hearts cannot apprehend that they are imaginatively thinking hearts, because we have so long been told that the mind thinks and the heart feels and that imagination leads us astray from both.

— JAMES HILLMAN

Emerging research, limited as it is, has begun to foster the reclamation of our hearts as organs of perception and communication. This research has, in general, focused in two areas: our internal world (our physiology), primarily in the context of helping to maintain health and in the understanding of a number of disease conditions, and our external world, specifically as it relates to our interactions with other people. Most of the research has begun by creating what a number of researchers call a state of coherence or entrainment.

HEART COHERENCE

Many of the studies conducted on the heart as an organ of perception and communication have focused on what happens when the heart's electromagnetic field is intentionally altered when a person shifts attention from linear, analytical processing (thoughts) to sensory stimuli, whether internal (listening to the heartbeat) or external (noticing how something looks, sounds, feels, or smells, for example). Researchers John and Beatrice Lacy comment, "The intention to note and detect external stimuli results in slowing of the heart. [This can be called the] bradycardia of attention."[3]

You can get a sense of this dynamic by sitting comfortably and looking at something that attracts your attention. Just let yourself look at it a moment, noticing its shape and colors. Then, let yourself notice how it *feels* to you. At that exact instant your entire physiological functioning will alter in a very noticeable manner. (But for it to happen you have to pay attention to the thing you are focused on, not the alteration you are expecting.)

This shift in the focus of awareness, from thinking to external sensory perception, significantly modifies and slows the duration of the cardiac cycle, producing a transformational cascade that affects all physiological and cognitive functioning. Simple attention to these external stimuli is sufficient. There need be no physical activity in response. Unlike linear, mental functioning, such as that required for mathematical calculations, there is no acceleration of heartbeat when focusing on external stimuli.

The immediate alteration in heart function that occurs with this shift in attention sends specific messages to the sensory-detecting areas of the brain and acts to facilitate—to enhance—these sensory perceptions. And the enhanced perception that comes with heart-focused perception does not habituate—in other words, perceived external events remain fresh and new each time this kind of dynamic is experienced.

This attention to the environment

whether internal or external

leads to a sympathetic-like dilation of the eyes, which become soft-focused instead of pin-point focused, with increased peripheral vision, at the same time that the heart slows—a parasympathetic activity. (Oversimply, the sympathetic part of the nervous system is concerned with flight or fight, the parasympathetic with rest and ease.) This indicates that both systems are at work but in a uniquely balanced manner.

soft-focused eyes and bodily relaxation

increases as the attention–interest value of a thing increases. The more interesting it is, the more this physiological state is enhanced.

The shift in cardiac function that occurs when one views external visual stimuli does not depend on the pleasantness or unpleasantness of what is viewed, but rather on its complexity, potency, one's personal evaluation of its nature, and its activity. These are common dimensions of meaning in a thing, along with novelty, surprisingness, and puzzlingness. The more meaning inherent in a thing, the more interesting it becomes and the greater the number of physiological alterations that occur. And these alterations are always accompanied by softer-focused eyes and a slowing down and relaxing of the body.

it is by this that you can recognize
the state of being

William Libby remarks, "An interesting, attention-getting stimulus, whether simple or complex, whether conveying a sense of activity and strength, or of passivity and weakness, evokes an autonomic response-pattern characterized by pupillary dilation and cardiac deceleration."[4]

Mental activities cause an almost immediate cessation of these physiological dynamics, with concomitant increases in heart rate and pupillary constriction. Any internal manipulation of symbolic information results in cardiac acceleration, an increase in sympathetic nervous system activity, and pupil constriction. So, too, does any verbalizing, or any requirement to store, manipulate, and retrieve symbolic information.

linear thinking breaks state

This shift in information processing and heart function initiates the beginning of what researcher Rollin McCraty calls a *state of coherence*. "It is the rhythm of the heart," he notes, "that sets the beat for the entire system. The heart's rhythmic beat influences brain processes that control the autonomic nervous system, cognitive function and emotions."[5] Coherence, he goes on to say, "is the harmonious cooperation, and order among the subsystems of a larger system that allows for the emergence of more complex functions. [It is used] to describe more ordered mental and emotional processes as well as more ordered and harmonious interactions among various physiological systems. [It] embraces many other terms that are used to describe specific functional modes, such as synchronization, entrainment, and resonance."[6]

In deepening this shift to coherence, most heart researchers emphasize a focus on personal emotional state as well as detection of external stimuli. Many ask study participants to intentionally generate the emotions of caring and affection.

Just as the communications embedded within the electromagnetic field of an organ or organism are experienced as emotions, if new emotions are intentionally created through conscious decision, they alter the form of the electromagnetic field, becoming embedded as new communications that then affect physiology.

These intentionally created emotional states initiate a repatterning of the heart's electromagnetic field, encoding new information. And this new information is used by the heart—or whichever other organism or organ it is directed toward—to alter its functioning. The heart's basic rhythm, McCraty reports, "is modified by the autonomic nervous system which is, in turn, modified by how we mentally or emotionally perceive events in the moment. . . Our emotions are reflected in the patterns of our heart rhythms. These changing rhythms appear to be modulating the field produced by the heart, similar to how a radio wave is modulated so that the music we hear can be broadcast."[7] Heart researcher Valerie Hunt clarifies, "Every experience has concomitant emotions, and every emotion temporarily restructures the field."[8]

Heart coherence begins when the location of consciousness is shifted from the brain to the heart, either through focus on the heart itself or on external sensory cues and how they feel.

[Psychology] has stumbled into the heart without a philosophy of its thought.

— JAMES HILLMAN

The heart is a tightly interconnected part of an oscillating, nonlinear neuronal network that is always processing electromagnetic waves within which information is encoded. During coherence, these interconnected networks couple with one another and begin working as one synchronized system.

When linear systems lock or couple together, the resultant patterns represent a simple mixture of the two systems. But when nonlinear systems, like the biological oscillators in our bodies, synchronize to a common frequency, the combined system resolves around a single oscillation. The difference between the oscillation frequency of the two (or more) systems begins to move toward zero. Unlike linear oscillators, when synchronized, nonlinear oscillators essentially become one oscillating pattern, in the waves of which ride information about all the nonlinear oscillators that have synchronized. This combination of two (or more) nonlinear oscillators has impacts; at its simplest, the amplitude of the combined waveform is significantly larger than that of either oscillator alone. This gives the coherent signal much more depth and power.

The electrical system of the body, produced by the body's natural oscillators, forms a coupled system of long evolutionary design with elegant feedback mechanisms among all the oscillators. When any one of these oscillators becomes the focus of consciousness, the other systems begin to entrain with it and boost its power. (The Chinese practice of *qigong,* used by Falun Gong adherents, focuses on the gastrointestinal (GI) tract as the primary locus of consciousness. The GI tract has its own extensive, elegant, and separate nervous system. Other practices, such as Heartmath training, focus on the heart.)

Against the background of the normal noise of the body, the electromagnetic field that emerges during coherence is highly detectable by the body's cells and organs, which entrain with it, amplify its signal, and use it to alter cellular and organ functioning.

Wide-ranging physiological impacts begin at the moment of coherence. When a person begins focusing on the heart, allowing him or her to immerse perceptions in its functioning, the coherence or synchronization that occurs begins in the heart. (Focus on the GI tract initiates these changes in that system.) Heart rhythms begin to take on a smooth, sine-wave-like pattern as all of the heart's electromagnetic frequencies start to synchronize. Normally, when our consciousness is phase-locked with the brain, the other biological oscillators in the body begin entraining with it. The result is much less coherent, because we seem evolutionarily designed to let the heart, the most powerful oscillator, be the primary system to which the others normally entrain. This coherent heart rhythm immediately begins to affect reticular neuronal activity.

The reticular neuronal network affects physiological functions, including respiration, somatomotor systems, and cortical activity. As the heart becomes more coherent, respiration, somatomotor systems, and cortical activity begin to entrain to the coherent heart rhythms. The three branches of the autonomic nervous system—sympathetic, parasympathetic, and enteric (the GI tract)—also begin to synchronize with this more coherent heart rhythm or wave pattern. Overall physiological functioning begins to be dominated by the parasympathetic, rather than the sympathetic (flight or fight). Sympathetic tone decreases; the body relaxes. There is a functional reorganization of autonomic balance. The respiratory system, at this point, begins to phase-lock to heart rhythms. Eventually, the heart, brain, and GI tract all couple together and demonstrate frequency-locking. Their oscillations shift to a frequency range that is the same for all three and the overall amplitude increases.

As coherence begins and deepens, the entire hormonal cascade of the body alters. This hormonal shift is initiated by the heart making and releasing significantly different amounts of its hormones and neurochemicals. As only one example: At coherence there is an average of 23 percent reduction in cortisol production (a stress hormone with negative impacts on immune function, memory and hippocampal function, and glucose utilization) and a 100 percent increase in DHEA production (an adrenal gland hormone essential in tissue repair, insulin sensitivity, sense of well-being, and sexual hormone production).

During heart coherence, ANF-induced alterations immediately occur at multiple target sites throughout the body: kidneys, adrenal glands, immune system, brain, posterior pituitary gland, pineal gland, hypothalamus, lung, liver, ciliary body (which secrets the lymphlike aqueous humor of the eye), and small intestine. ANF alterations immediately read-just the complex balance of our whole, interconnected physiology. Blood pressure lowers, muscle cells throughout the vascular system relax, eye function alters.

Atrial naturetic factor binds to a number of sites in the eye, affecting ocular pressure and eye focus. With this alteration in ANF and its imme-diate impacts on the eye, the eyes become soft-focused, peripheral vision is enhanced.

In addition, coherence affects levels of other heart hormones, brain naturetic factor and C-type naturetic peptide, which also shift physiology and brain function, especially in the hypothalamus, adrenal glands, and pituitary gland. Secretion of beta-amyloid precursor protein increases, protecting neurons from stressors throughout the brain and especially in the hippocampus. Changes in levels of ANF, CNP, and BNF directly affect the hippocampus, enhancing its functioning. These changes increase dopamine production in the heart, improving the transfer of information from neuron to neuron in both heart and brain.

HEART-BRAIN ENTRAINMENT

When the brain entrains to the heart, connectivity increases between brain and body. Conversely, the location of consciousness in the brain leads to an increased disconnection between brain and body. When one shifts into heart-oriented cognition, mental dialogue is reduced.

> *One becomes aware of an inner electrical equilibrium.*
> — WILLIAM TILLER

Sympathetic and parasympathetic nerve pathways and the barore-ceptor system directly link the heart and brain, allowing communica-tions and information to flow freely. Messages flowing from the heart to the brain during this shift to coherence significantly alter the brain's functioning, especially in the cortex, which profoundly affects percep-tion and learning.

> *The major centers of the body containing biological oscillators can act as coupled electrical oscillators. These oscillators can be brought in to synchronized modes of operation through mental and emo-*

*tional self-control and the effects on the body of such synchroniza-
tion are correlated with significant shifts in perception.*

— WILLIAM TILLER

Thus, a new mode of cognition is activated: the holistic/intuitive/depth mode.

Heart researcher McCraty comments, "[heart entrainment] leads to increased self-management of one's mental and emotional states that automatically manifests as more highly ordered physiological states that affect the functioning of the whole body, including the brain. The practitioners of these heart focus techniques report an increased intuitive awareness and more efficient decision-making capability that is beyond their normal capacity from the mind and brain alone."[9]

Shifting the focus of consciousness to the heart—and away from the forebrain—results in entrainment of large populations of cells in the forebrain to cardiac functioning (rather than vice versa). These populations of forebrain cells begin oscillating to the rhythms produced by the heart, and the perception of those populations of cells, the kinds of information they begin to process during entrainment, is very different from what they process when entrainment is not occurring.

The human brain operates in a state that is far from equilibrium; it, like the heart, is a complex, nonlinear oscillator. Every day, there is an incessant stream of incoming data—material to "think" about. These incoming signals cause the system to constantly shift from one state to another in response to the incoming signals. The system constantly wobbles in and out of dynamic equilibrium, reestablishing a new homeodynamic every time it is perturbed. The neurons in the brain are nonlinear, oscillators themselves, and can be influenced by extremely weak perturbations. They are very sensitive to such perturbations, for they, like all nonlinear oscillators, use stochastic resonance to boost signal strength. A shift in the heart's electromagnetic field is a perturbation that the brain has been evolutionarily intended to respond to. And when the heart goes coherent, the brain immediately begins to respond.

Coordinated interactions across extracellular space lead to long-range, coordinated dynamics of heart and brain function during heart/brain entrainment. When brain neurons entrain to the heart's ECG activity, the timing of neuronal firings alters, and research shows that the timing of neuronal firing conveys several times more information than the firing count. Analysis of electroencephalogram readings shows that the heart's signals are strongest in the occipital (posterior) regions of the brain and the right anterior (front) sections of the brain. The brain's alpha rhythms also synchronize to the heart, and their amplitude lowers when they do so. The brain's alpha rhythms are the fastest of the brain's electromagnetic waves. Their amplitude is lower when brain arousal is lower or when a person concentrates on external sensory phenomena rather than on abstract analytical or symbolic thoughts.

After heart/brain entrainment, when a combination of both heart and brain waves are taken by electrocardiogram, what is seen is that the brain waves ride on top of the heart waves. Not only are they oscillating together, the brain's wave patterns are, in fact, embedded within the larger field of the heart.

Hippocampal activity increases considerably when cognition is shifted to the heart, heart coherence occurs, and the brain entrains to the heart. Focusing on external sensory cues activates hippocampal functions, since all the sensory systems of our bodies converge in the hippocampus. The increased demand on hippocampal function stimulates stem cells to congregate in the hippocampus and form new neurons and neuronal complexes. The reduced cortisol production that occurs during heart coherence directly enhances hippocampal activity as well. The hippocampus, in other words, comes strongly online. It begins sifting the electromagnetic fields the heart is detecting for embedded patterns of information, eliciting meaning from background information. The hippocampus then sends information about those meanings to the neocortex, where it is encoded as memories. The more that sensory focus is on external environments, the more activated the hippocampus and its analysis of meaning becomes.

Shifting attention to any particular organ—in this case, the heart—increases registration of the feedback from that organ in the brain. This increase is measurable in electroencephalogram patterns. The shift to heart awareness initiates an alteration in body functioning via physiological mechanisms that operate through neural registration of organ feedback on the brain.

This kind of synchronization does not occur spontaneously, unless people habituate heart-focused perception. Since we have been habituated to the analytical mode of cognition through our schooling, taught to locate our consciousness in the brain and not the heart, this type of entrainment must be consciously practiced. (For most of us, heart-focused perception is not a natural mode of processing information, though it was for ancient peoples and sometimes still is for indigenous cultures.) Even though the brain entrains with the heart through heart-focused techniques, the brain tends to wander in and out of entrainment. Because of the brain's long use as the dominant mode of cognition, this entrainment is not permanent. Practice in entrainment helps the brain and any other system to maintain synchronization for longer and longer periods of time.

Impacts on Health and Disease

The heart is the most powerful oscillator in the body and its behavior is naturally nonlinear and irregular. One measure of the irregular, nonlinear activity of the heart is called heart rate variability or HRV. The resting heart, instead of beating regularly, engages in continual, spontaneous fluctuations. The heartbeat in young, healthy people is highly irregular. But heart beating patterns tend to become very regular and predictable as people get older or as their hearts become diseased. The greater the HRV, the more complex the heart's beating patterns are and the healthier the heart is.

> Complexity here refers specifically to a multiscale, fractal-type variability in structure or function. Many disease states are marked by less complex dynamics than those observed under healthy conditions. This decomplexification of systems with disease appears to be a common feature of many pathologies, as well as of aging. When physiological systems become less complex, their information content is degraded. As a result they are less adaptable and less able to cope with the exigencies of a constantly changing environment. To generate information a system must be capable of behaving in an unpredictable fashion. . . Certain pathologies are marked by a breakdown of this long-range organization property, producing an uncorrelated randomness similar to white noise.
>
> — Ary Goldberger

What is especially telling is that when the heart is entrained to the brain's oscillating wave-form, rather than vice versa, the heart begins to, over time, lose coherence. The more the heart entrains to the brain, and the longer it does so, the less it displays a variable HRV, the less fractal its processes are, and the more regular it is. It is, in fact, entraining to a linear rather than a nonlinear orientation. It is not surprising then that our culture's focus on a type of schooling that develops the brain to the exclusion of the heart, that fosters thinking instead of feeling, detachment instead of empathy, leads to disease. Heart disease is the number-one killer in the United States.

When *any* system begins to lose this *dynamical-chaos* aspect of its functioning and becomes more predictable, it begins to lose elegance of function. It, in fact, becomes diseased. Heart disease is always accompanied by an increasing loss of nonlinearity of the heart. The more predictable and regular the heart becomes, the more diseased it is. Loss of heart rate variability, for instance, occurs in multiple sclerosis, fetal distress, aging, and congestive heart disease. To be healthy, the heart must remain in a highly unstable state of dynamic equilibrium.

Given all this, it is not surprising that unhealthy emotional states—major depression and panic disorders, for example—correlate with changes in HRV as well as alterations in the power spectral density of the heart. (Power spectral density refers to the *range* and *number* of electromagnetic waves produced by the heart.)

During major depression and panic disorder, as in many pathological heart conditions, the heart's electromagnetic spectrum begins to show a narrower range, and beating patterns again become very regular. This narrowing and increase in regularity also show direct impacts in the sympathetic and parasympathetic nervous systems. Sympathetic nervous system activity and tone tend to increase, the parasympathetic to decrease. These are all signs of increasing heart disease, as a disordered heart cannot produce the extreme variability and flexibility that is normal in the healthy heart. Because emotional experience comes, in part, from the electromagnetic field of the heart, a disordered, narrow, noncomplex electromagnetic field will produce emotional experiences, like depression and panic attacks, that are themselves disordered, narrow, and restricted in scope.

In many pathological conditions, the heart's electrophysiologic system acts as if it were coupling itself to multiple oscillatory systems on a permanent basis. In other words, it behaves as if it can't make up its mind, and its cells no longer beat as one unified group. Instead, the group begins to split (broken-hearted), pulled this way and that by different outside oscillating attractors. Holding the consciousness to one state of being, the verbal/intellectual/analytical mode of cognition, of necessity produces a diminished heart function, a shallower mix of emotional states, and an impaired ability to respond to embedded meanings and communications from the environment and from the self.

Conversely, increasing heart coherence and heart/brain entrainment has shown a great many positive health effects. Increased heart coherence boosts the body's production of immunoglobulin A, a naturally occurring compound that protects the body's mucous membranes and helps prevent infections. Increased heart coherence and heart/brain entrainment also produces improvements in disorders such as arrhythmia, mitral valve prolapse, congestive heart failure, asthma, diabetes, fatigue, autoimmune conditions, autonomic exhaustion, anxiety, depression, AIDS, and post-traumatic stress disorder. In general, in many diseases, overall healing rates are enhanced.

One specific treatment intervention study, for example, found that high blood pressure can be significantly lowered within six months—without the use of medication—*if heart coherence is reestablished.* And as heart/brain synchronization occurs, people experience less anxiety, depression, and stress overall.

Lack of cognitive focus on the body (habituation to the verbal/intellec-tual/analytical mode of cognition) results in disconnection and increased dis-order in organ function—and is the foundation of many diseases, including heart disease. When attention is focused on different sensory cues (e.g., heartbeat, respiration, external visual stimuli) physiological function shifts significantly and becomes more healthy. It becomes even more healthy when specific kinds of emotions are activated: feelings of caring, love, and appre-ciation enhance internal coherence. The more confused, angry, or frustrated a person becomes, the more incoherent their heart's electromagnetic field.

> *I declare that a meal prepared by a person who loves you will do you more good than any average cooking, and on the other side of it a person who dislikes you is bound to get that dislike into your food, without intending to.*
>
> — LUTHER BURBANK

In the healthy heart, the varied and complex emotional mix we experi-ence each day—generated by contact with our internal and external worlds—produces a range of heart rate patterns that is nonlinear and con-stantly shifting. Communications are embedded within these shifting mixes and patterns, communications from and to our bodies, our loved ones, the world at large. The narrower the range of the electromagnetic spectrum, the more regular the beating patterns of the heart and the less "hearty" we become.

HEART COMMUNICATION WITH THE EXTERNAL WORLD

Biological fields, as Renee Levi comments, are "composed of vibrations that are organized, not random, and have the capacity to selectively react, interact, and transact internally and with other fields."[10] "Our body and brain, Joseph Chilton Pearce remarks, "form an intricate web of coherent frequencies organized to translate other frequencies and nestled within a nested hierarchy of universal frequencies."[11]

Living organisms, including people, exchange electromagnetic energy through contact between their fields, and this electromagnetic energy carries information in much the same way radio transmitters and receivers carry music. When people or other living organisms touch, a subtle but highly complex exchange of information occurs via their electromagnetic fields. Refined measurements reveal that there is an energy exchange between people, carried through the electromagnetic field of the heart, that while strongest with touch and up to 18 inches away, can still be measured (with instruments) when they are five feet apart.

> Though of course, our (technological) ability to measure electromagnetic radiation is very crude. Electromagnetic signals from living organisms, just like radio waves, continue outward indefinitely.

Thus energy, encoded with information, is transferred from one electromagnetic field to another. In response to the information it receives, the heart alters its functioning and encodes in its fields, on a constantly shifting basis, its responses. Those responses can, in turn, alter the electromagnetic fields of whatever living organisms the heart is engaged with—for this is a living, ever-shifting dialogue.

The heart generates the strongest electromagnetic field of the body, and this field becomes more coherent as consciousness shifts from the brain to the heart. This coherence significantly contributes to the informational exchange that occurs during contact between different electromagnetic fields. The more coherent the field, the more potent the informational exchange.

A coherent heart affects the brain wave pattern not only of the person achieving coherence, but also of any person with whom it comes into contact. While direct skin-to-skin contact has the greatest effect on brain function, mere proximity elicits changes. A sender's coherent heart-field is measurable not only in a receiving person's electroencephalogram, but also in his or her entire electromagnetic field.

When people touch or are in close proximity, a transference of their heart's electromagnetic energy occurs, and the two fields begin to entrain or resonate with each other. The result is a combined wave created by a combination of the original waves. This combined wave has the same frequency as the original waves but an increased amplitude. Both its power and depth are increased.

The signal of transfer is sometimes, but not always, detected as flowing in both directions; this depends to a great extent on the context of the transfer and the orientation of the sender. When a person projects a heart-coherent field filled with caring, love and attention, living organisms respond to the information in the field by becoming more responsive, open, affectionate, animated, and closely connected.

The importance of caring on outcomes in healing has been stressed in a great many cultures and types of healing professions. Healing practitioners that consciously produce coherence in the electromagnetic field of their hearts create a field that can be detected by other living systems and their biological tissues. This field is then amplified and used by the organism detecting it to shift biological function. When these loving, practitioner-generated fields are detected and (naturally) amplified by ill people, healing rates of wounds are increased, pain decreases, hemoglobin levels shift, DNA alters, and new psychological states manifest.

Thus, the best outcomes are dependent on the state of mind of the healer. Extreme importance should be attached to the kind of intention a practitioner has as he or she works. The more caring the practitioner, the more coherence there will be in their electromagnetic field and the better the healing will be.

When we are cared for or care for others, the heart releases an entirely different cascade of hormonal and neurotransmitter substances than it does in other, less hopeful, circumstances. Falling in love causes a tremendous expansion of the heart, a flood of DHEA and testosterone throughout the heart and body, and a flow of other hormones, such as dopamine, all of which affect adrenal, hypothalamus, and pituitary hormone output.

More Immunoglobulin A, or IgA, is also released, stimulating the health and immune action of mucous membrane systems throughout the body.

The receiver's receptivity to the practitioner's heart-field also plays a part in the outcome. The more open he or she is to receiving caring, the more he or she will entrain with an external electromagnetic field. However, the elegance of the practitioner in creating and directing a coherent electromagnetic field to the patient is of more importance than the sufferer's receptivity. In addition, the practitioner-generated field must be continually adjusted.

Because the heart's electromagnetic field is nonlinear, healers can alter the makeup of the field through a constantly shifting perception of the patient. As the healer shifts toward coherence, not surprisingly, there is an alteration in his or her own cortical function. At this point, personal perception also alters considerably. The healer's cognition is, as McCraty puts it, "dramatically changed."[12] This altered perception is by nature extremely sensitive to the fabric of external electromagnetic fields and the information contained within them. As the practitioner's perception and their facility in using it deepens, it is possible to use it in a highly directed fashion to extract more meaning from the patient and his or her interior world. As the patient's electromagnetic field alters, as it will from moment to moment throughout the process, the kind of caring, attention, and love the practitioner sends and where it is directed can be adjusted, making it more highly sophisticated in its impacts. Because the healer's electromagnetic field is so personally directed and shaped to fit the unique needs and electromagnetic field of the patient, the patient's sensitivity to the process increases the more it occurs. *Anyone* can, and will, respond with significant shifts in their electromagnetic field if the practitioner's technique is elegant enough.

(If the practitioner entrains him-or herself to the patient's ECG or EEG, their heart can take on the disease patterns in the other person—beat and EEG pattern, and so on. Self-reflection will show the practitioner the pattern of disease in the patient, and by altering their own pattern back toward health, the practitioner can determine the processes, the steps necessary to produce health in the patient. But beyond this, the patient, in a state of synchronization, will tend to "follow" the leads embedded in the practitioner's electromagnetic field, moving toward health.)

The more accustomed people become to responding to coherent electromagnetic fields generated through a practitioner's heart, the more rapidly they are able to physiologically respond when they detect a coherent electromagnetic field. The more interaction two living organisms have, the more imprinting that occurs on their hearts, the more alteration there is in their electromagnetic fields, the more shifts that occur in heart function. Because this element of healing is almost absent in conventional, techno-logical medicine, patients are not used to responding to coherent electro-magnetic fields as part of their healing. In fact, the electromagnetic field of most medical healers is extremely incoherent, since they have been trained to use their brains to the exclusion of their hearts. The ill are immersed in incoherent electromagnetic fields throughout their healing process in hos-pitals, which, in and of itself, is a strong contributing element to the kinds of outcomes hospitals and physicians produce.

We have all some electrical and magnetic forces within us; and we put forth, like the magnet itself, an attractive or repulsive power, as we come in contact with something similar or dissimilar.

— GOETHE

BEYOND PEOPLE

But heart-centered communication is not limited merely to the body and other people. The heart, through its electromagnetic field, continually senses electromagnetic patterns from its environment and works to decode the information contained within them. Perturbations that can affect the dynamic equilibrium of the whole oscillating, self-organized systems that we know as ourselves come not only from within, but also from without.

The tendency to focus emerging research solely on the interrelation-ship of electromagnetic fields to internal health or the interactions that occur between people are an expression of our *anthropocentricism,* our human-centeredness. This narrowing of the understanding of electromag-netic fields is a prime example of our application of a hierarchy of values that places human beings at the top and our belief that the rest of the world is filled with things put here for our use—a reflection of our belief that we are the most important organisms on the planet and the only organisms with intelligence and soul. But all living organisms produce electromagnetic fields, all encode information, and all merged electro-magnetic fields exchange information. The Earth itself is a living organism that produces electromagnetic fields filled with information. We are affected by the information encoded in these fields just by living on the

Earth. Many periodic rhythms in our bodies are a function of our entrainment to the oscillations of the electromagnetic field of the Earth. Circadian rhythms are the reaction of living organisms to periodic electromagnetic fluctuations in the environment.

If all external inputs are severed (by putting people in space or in a sealed, enclosed environment, for example), the rhythms continue in our bodies, but in a very different manner. These rhythms *are* generated internally in all living organisms but their periodicity—their timing—is shifted by the electromagnetic fields in which they are nestled.

When a human is placed in an environment in which there are no time cues, the daily activity cycle gradually lengthens. This means that our normal 24-hour day involves an external entrainment of our endogenous circadian generators. . . body temperature and autonomic functions adapt also, but more slowly. The biological importance and omnipresence of autogenous rhythmicity have been largely underestimated. Such periodicities must be considered as a phylogenetic adaptive mechanism to the time structure of our environment, which has been maintained genetically.

— G. SIEGEL

Oscillating external electromagnetic fields can entrain or phase-lock heart cells so that the organism that we know as ourself moves into synchronicity with those electromagnetic fields. We are, in fact, supremely able to perceive and be affected by extremely weak electromagnetic fields from the environment.

There is no fundamental lower limit with respect to the magnitude of the perturbation that is still capable of influencing a nonlinear oscillator.

— PAUL GAILEY

There is a tendency among many reductionists to *mechanomorphize*, to project onto the world around them the belief that there is no intelligence in anything other than human beings, that life is merely the result of mechanical forces. Thus, when these kinds of researchers examine Nature, they tend to see and find what they already believe is there. However, all life gives off electromagnetic fields, all life has been bathed in such fields for the nearly 4 billion years that life has existed on this Earth. These electromagnetic fields are not merely the unconscious expressions of mechanical functioning. Living organisms, over long evolutionary time, have learned to use these fields as a communication medium, to intentionally insert information into them.

The constantly interblending flow of information-loaded electromagnetic fields is part of the communication dynamic of living organisms within ecosystems, an aspect of their coevolutionary bonding. Electromagnetic fields are used not only for supporting the integrity of the organism—for strengthening physical structure and healing it when damaged—but also for deterring hostile organisms (like the unfriendly, defensive fields that an attack dog expresses even without growling). Perhaps even more important, these fields are used to strengthen cooperative interactions among organisms within ecosystems. Because of our anthropocentrism, this is more obvious within small organism groupings, such as cells within bodies or members of human families, whose interwoven loving bonds represent the long-term intermingling of supportive, cooperative, coevolutionary electromagnetic fields that are continually embedded with complex information designed to enhance those connections. But such families and their individuals are nested within and encounter a wide variety of such fields, including fields from plants.

Plants, like all living organisms, generate and respond to electromagnetic waves. They use a great many internal electromagnetic communications, just as we do, for healing and for normal physiological functioning. For like us, they are composed of millions upon millions of cells. But what is less well known is that like us, plants also have very sophisticated central nervous systems.

The characteristics of conduction in the plant nerve are in every way similar to those in animal nerve.

— JAGADIS CHANDRA BOSE

In many respects, plant nervous systems are nearly as sophisticated as our own, and in some plants, nearly as rapid in their actions. Plant nervous systems possess synapses, just as our brains do, and they make and use neurotransmitters that are molecularly identical to those that are found in our brains. They use these neurotransmitters to facilitate the function of their central nervous system, just as we do.

Plant nervous systems perform many of the same duties ours do—they help process, decipher, and coordinate external and internal impulses to maintain the functioning of the organism. And a major element of this functioning is their recognition of signals, their decoding of meaning, and their crafting of responses. The great Indian researcher Jagadis Chandra Bose conducted perhaps the most sophisticated exploration to date into the nervous systems of plants. In his book, *The Nervous Mechanism of Plants,* he comments,

> In light of the results summarized in this chapter, it can no longer be doubted that plants, at any rate vascular plants, possess a well-defined nervous system.
>
> It has been demonstrated that excitation is conducted by the phloem of the vascular bundle, and that conduction in this tissue can be modified experimentally by the same means as it is in animal nerve. The conducted excitation may, therefore, be justly spoken of as nervous impulse and the conducting tissue as nerve.
>
> It has been further shown that, as in the animal, it is possible to distinguish sensory or afferent and motor or efferent impulses, and to trace the transformation of the one into the other in a reflex arc. The observations involve the conception of some kind of nerve center.[13]

Plant nervous systems are as highly sensitive to electromagnetic fields as ours. This is necessarily so, since they use the electromagnetic energy of the sun in photosynthesis. They emerged as an ecological expression of Earth specifically to work with the electromagnetic spectrum. But the range of their sensitivity goes far beyond the spectrum of visible light. They can, in fact, detect and respond to broadband electromagnetic signals, as can all organisms.

> *There is no life-reaction in even the highest animal which has not been foreshadowed in the life of the plant. . . The barriers which seemed to separate kindred phenomena will be found to have vanished, the plant and the animal appearing as a multiform unity in a single ocean of being. In this vision of truth the final mystery of things will by no means be lessened, but greatly deepened. [For] that vision crushes out of [Man] all self-sufficiency, all that kept him unconscious of the great pulse that beats through the universe.*
>
> — JAGADIS CHANDRA BOSE

And we, like plants, are evolutionarily designed to encounter such fields, just as the generators of those fields are designed to encounter us. The meanings embedded within those fields, experienced by us as emotions, affect the heart's rate, hormonal cascade, pressure waves, and neurochemical activity. Directed emotions—intentional, informational electromagnetic embeds sent outward—affect those external electromagnetic fields in turn. Through such directed communication and perception, a living dialogue occurs between us and the world.

Such interchanges are a part of what it means for us to be human and have been a part of our interaction with our environment since we emerged out of the living field of this planet. But without a flexible heart, they cannot be perceived.

Only to him who stands where the barley stands and listens well will it speak, and tell, for his sake, what man is.

— MASANOBU FUKUOKA

CHAPTER SIX

THE SPIRITUAL HEART

AISTHESIS

*Sensorially disconnected from their theoretically evolved informa-
tion, scientists discern no need on their part to suggest any educational reforms to correct the misconceiving that science has tolerated
for half a millennium.*

— BUCKMINSTER FULLER

The heart brings us authentic tidings of invisible things.

— JAMES HILLMAN

The mystery of life isn't a problem to solve but a reality to experience.

— FRANK HERBERT

*When microscopic vision fails we still have to explore the realm of the
invisible.*

— JAGADIS CHANDRA BOSE

Even though the foregoing five chapters are a good metaphor, they are only a metaphor.

literally, and not metaphorically, they are a metaphor
this kind of reductionism is never real
these metaphors are only a way to think about something
that is meant to be experienced

The important thing is not that the heart is an intricate pump or even that it creates electromagnetic fields that other people can feel or through which we can communicate. The important thing is that we exist, immersed in living fields of communication, all of which are imbued with meaning, generated by intelligent life forms, and flow from and to us from the moment the cells of our bodies self-organize into the unique identities that we know as ourselves.

We experience these communications not as lines of words on a page, but as multivalued, complex exchanges of intentionality, touches of the living intelligence of life forms to which we are kin. They are exchanges of the *qualities* inherent in living organisms, not quantities of mechanical forces. We feel the touch of the world upon us, and those millions of unique touches hold within them specific meanings, sent to us from the heart of the world and from the heart of the living beings with which we inhabit this world. This interchange changes the quality of our lives and reminds us that we are never alone. We are one organism among many, one ensouled form among a multitude.

The modern world's belated (and very mild) recognition that the heart is an organ of perception is not new, of course. By locating our consciousness in only one biological oscillator, the brain, we blinded ourselves to perceptions that have been common to human beings since they emerged from this Earth. In gaining a reductionist understanding of the world, we lost touch with the essential nature of the Earth and ourselves.

As Earth leaves,
we remain,
stones,
not plants,
not green,
lonely pebbles
scattered
on an empty street.

The Greeks had a word for the heart's ability to perceive meaning from the world: *aisthesis*. "In Aristotelian psychology," James Hillman notes, "the organ of *aisthesis* is the heart; passages from all the sense organs run to it; there the soul is 'set on fire.' Its thought is innately aesthetic and sensately linked with the world."[1]

Aisthesis denotes the moment in which a flow of life force, imbued with communications, moves from one living organism to another. The word literally means "to breathe in." It is a taking in of the world, a taking in of soulful communications that arise from the living phenomena in that world. The ancient Greeks knew that this moment of recognition was usually accompanied by a gasp, a breathing in. Something from outside enters inside us, something with tremendous impact, something that causes an immediate inspiration. Often overlooked is the crucial understanding that at the same time, the world takes us in too—we are breathed in as well. When we experience this sharing of soul essence, we have a direct experience that we are not alone in the world. We experience the truth that we live in a world of ensouled phenomena, companioned by many forms of intelligence and awareness, many of whom care enough for us to share this intimate exchange.

Once we were convinced that there was no intelligence, no living soulful force, in Nature, once we were convinced that the heart was nothing more than a pump, we began to lose touch with our innate capacity for engaging in aisthesis, for feeling the touch of the living world upon us, interpreting what that touch means, and sending out from ourselves a response in turn.

Scientists have engaged in a particular form of imperialism. They have stolen from all of us the historical recognition of the heart as an organ of perception and substituted instead a mechanical heart and the belief that the brain is the only organ capable of thought. This colonization of the soul has had profound repercussions.

The man of science, who is not seeking for expression but for a fact to be expressed merely, studies nature as a dead language.
— HENRY DAVID THOREAU

Though the brain is important, it is merely an organic computer, useful for processing data and acting as a clearing station for central nervous system functioning. Unlike the heart, with its connected empathic perceptions the brain has no inherent moral nature. The continual training of children in a system of perception that is amoral leads to behaviors in adults that have no moral basis.

> *Thomas Huxley,*
> *Darwin's strongest defender,*
> *observed*
> *that "No rational man,*
> *cognizant of the facts,*
> *believes that the Negro*
> *is the equal*
> *still less the superior,*
> *of the white man."*

> *The assertion*
> *that the degree of rational "thinking"*
> *of any species*
> *is illustrative of its position*
> *on the ladder of evolutionary hierarchy*
> *is only a decision*
> *by an organism*
> *(and specific people)*
> *with a vested interest*
> *in what is being decided.*

> *What does*
>   *a bristlecone pine*
>   *do*
>   *during six thousand years*
> *of life?*

> *What does*
>   *a blue whale*
>   *do*
>   *with the largest brain*
> *on Earth?*

The perceptions of the world that occur through the brain are themselves colored by the organ that is used for primary perception. The inherent linearity of the brain can only perceive the exterior world in a linear fashion, as a collection of externally situated, Euclidean objects.

Suppose that a scientist wants to understand nature. He may begin by studying a leaf, but as his investigation progresses down to the level of molecules, atoms, and the elementary particles, he loses sight of the original leaf.

— MASANOBU FUKUOKA

The linear brain cannot perceive wholes—or insides. And the more the brain is used as the primary organ of perception, the more life is reduced. It becomes merely an expression of mechanical forces with no intelligence or purpose, and all life forms are judged and valued depending on their capacity for this kind of analytical processing. Descartes' famous dictum, "I think, therefore I am," affirms as well its opposite: If you do not think, you are not.

What does education often do? It makes a straight-cut ditch of a free, meandering brook.

— HENRY DAVID THOREAU

People who intuitively know the importance of feelings, who use the heart as an organ of perception, receive little cultural support for this ancient mode of cognition. Their observations are routinely denigrated, and nonlinear thinkers have spent centuries attempting to prove that there is, in fact, another route to perception of the world than that offered up by the brain.

The present age has a bad habit of being abstruse in the sciences. We remove ourselves from common sense without opening up a higher one; we become transcendent, fantastic, fearful of intuitive perception in the real world, and when we wish to enter the practical realm, or need to, we suddenly turn atomistic and mechanical.

— GOETHE

Aisthesis happens to us still, although few of us consciously understand what is happening. On seeing the Grand Canyon or coming unexpectedly upon a beautiful, ancient tree in a forest, there is an immediate turning, a stopping, and then a gasp, a breathing in, as the power of the thing is felt. But it doesn't happen often, at least not continually, as it did for most of our species throughout most of our time on Earth. For such soul-sharing is impossible in a mechanistic universe that possesses no soul. The soul of a thing cannot leave its physical form and enter us if it possesses no soul in the first place. Nor, if there is nothing there, can we be "inspired" by the breathing in of the world.

Still, this basic experience—this aisthesis—has been at the root of human relationship with the world since our evolutionary expression out of the Earth. We are built to experience it, to be aware that each thing possesses a unique identity, its own particular *eachness*. We are made for the nature of each thing to pass into us through our hearts, which think about it, store memories about it, and engage in dialogue with it.

The more we know, the more mysterious it becomes that we can and do know both aught and naught. The number one a priori characteristic of the entirely mysterious life is awareness.

— BUCKMINSTER FULLER

Like all developed human skills, it takes years of exploring and experimenting with the perceptual capacities of the heart for it to become sophisticated, just as it takes years to develop fluency with verbal language. Unfortunately, because we are trained out of using our hearts as perceptual, thinking organs during our lengthy school years, in later life, if we try to begin using this ability, it is often a fumbling, awkward experience. Our heart intelligence is still working at a six-year-old level while our mind intelligence is absurdly further ahead.

Still, there is a great power in the world around us. It has not disappeared just because we no longer notice it. Redeveloping the capacity for heart-centered cognition can help each of us reclaim personal perception of the living and sacred intelligence within the world, within each particular thing. It moves us from a rational orientation in a dead, mechanized universe to one in which the unique perceptions of the heart are noticed and strengthened, to a deep experience of the living soulfulness of the world. As the process continues to deepen, it strengthens our spiritual sensitivity and, in the process, helps us gain a deeper understanding of our own sacredness. During this reclamation of our ability to *feel and think* with the heart, there is often a period when the accumulated scar tissue that covers most of our hearts begins to be stripped away. This allows the heart to become flexible again, once more enabling us to use it as an organ of perception.

The use of the heart as an organ of perception and communication, to weave us once again inextricably into the life web of the Earth, to gather knowledge from the heart of the world, and to help us live a whole and fulfilled life, to become who we are meant to be, is what the rest of this book is about.

The small ruby everyone wants has fallen out on the road.
Some think it is east of us, others west of us.

Some say, "among primitive earth rocks," others, "in the deep waters."

Kabir's instinct told him it was inside, and what it was worth,
and he wrapped it up carefully in his heart cloth.

— KABIR

DIASTOLE

GATHERING KNOWLEDGE FROM THE HEART OF THE WORLD

*There is one place
in all the Universe
that has been made
just for you.
And it is inside
your own feet.*

PROLOGUE TO PART TWO

It is the heart that always sees, before the head can see.
— THOMAS CARLYLE

You must have a feeling for the organism.
— BARBARA McCLINTOCK

YOU FIND SKUNK CABBAGE while walking deep in wetland bogs and shadowed forest. For it belongs to an ancient world, a world ancient long before humans walked or talked or breathed. You must wear boots when you go to find it, and dirty clothes. Skunk cabbage is not a plant for white shirts, not a plant for the fastidious.

we're going to get dirty on this one

The trees in such wetland forests stand mossy-sided, ragged-barked, limbs jagged-reaching, feet deep in wet earth and watery pools. Their canopy heavily shadows the stillness. The plants here don't like bright sun. They like it moist and dark and still.

You enter the deepening gloom knowing you are entering another world. The farther you walk in from the boundaries of the wetland, the more you travel back in time, to when horsetail towered over the land.

You'll find skunk cabbage then, if you're blessed, its leaves spreading out and up from the wetness in which it grows. The pools in which it lives are not stagnant. The water is constantly moving, interpenetrating everything. This wetness gleams on the plant's surfaces, a thin layer of living water.

You become aware of its leaves immediately. Green. A dark green with lighter hues merging into it across the face of the leaves. Its hooded flower is large, yellow, a knobby spike when the seeds form, like corn.

sort of

The plant and flower are some kind of huge throwback to a time when dinosaurs ruled the land. There is a silence that reigns where skunk cabbage grows. The mind quiets as it nears the plant, becoming more silent than even the stillness coming from the forest that surrounds you. You approach it slowly, reverently, and kneel at its side, the muddy, watery soil soaking through the knees of your pants. The water and mud begin their work immediately on any who seek this medicine.

Involved in ancient things, the plant takes a while to awaken to your presence. But you stay with it, asking for help, for its acquiescence in coming with you as medicine. Skunk cabbage is not like other plants that medicine seekers harvest. When you take the shovel and begin to dig, the plant resists the straight-forward approach. You must dig beside it. Not *under* it. A moat is the way to go.

If you place the blade of the shovel under the plant

everyone does this once

and press down, trying to lever the plant up, the shovel, and you, sink deeper into the muck. The plant remains unmoved. Unmoved by your entreaty. So you dig a moat.

As fast as you dig, the moat fills with water. And the plant's roots are not one shovel deep,

they do not make it easy

but one-and-a-half. So you must dig again, into the muddy water, striving to get the shovel through the tangled roots that interweave the soil in all such wetlands. The intertangled roots of these plant communities like ancient pools and shadowed forest.

But you persevere, and the shovel goes in. The water-logged soil makes a sucking sound as you lever it up and out of the ground. The mud is like glue and pulls tightly on the shovel, holds it in place, resists any attempt to pull it up and out. Wet land resists any removal of its soil. But you keep on.

The muddy glop of water and soil comes up finally, in a long, slow slurp as the ground lets go at last. The load is heavy as you lift it to the side. You turn the shovel to the side and the mud falls to the ground in a heavy plop. And then you turn back to the moat you are making, inserting the shovel back into its watery hole.

Once the moat encircles the plant, once you have deepened it to one-and-a-half shovel blades in depth, you must abandon the shovel. And then you must really *immerse* yourself into the experience, for only hands can do the work now.

Putting the shovel aside, you kneel in the muck, reaching with your hands into the moat, the muddy water sliding up your forearms. You have to feel with your fingers, wriggling them into the wet soil, under the plant. The plant roots whirl out from a small, almost inadequate, central bulb, each rootlet the size of an earthworm. The roots are segmented, like worms, and they hold on tightly, just like worms. They cannot be pulled out, for they pull back as strongly—meeting your pull, strength for strength. So you feel them out, one by one. Loosening them from the wet soil, removing the soil from around them.

The plant is huge, two to three feet high, spreading its giant leaves out perhaps three feet in diameter from the plant's center. So you work your way around it. On hands and knees first, then sitting, almost lying in the muck. You can't see what is happening; the muddy water has filled the moat completely. It must all be done by touch. But eventually,

eventually

the last rootlet is found, released from the soil, and you grasp the plant by the stalk just beneath the place at which the leaves begin to spread out. And you lift it out of the hole. The rootlets are covered with mud and are hard to make out. But the hole has immediately filled with muddy water, so you dunk the plant in the pool. Quickly, in and out. And again, perhaps two or three times. Most of the mud washes off as you do this, and then you can see her.

The root is huge and the rootlets like Medusa's hair, a swirling, moving set of tendrils wriggling out from the bottom of the plant. It almost seems as if they *are* the plant. The whole of your attention focuses upon them. You almost want to turn the plant upside down, as if the roots are the head, the leaves the root. So, eventually, you do. You feel yourself then looking at some ancient being, find yourself in some ancient story whose telling began long ago.

The roots are thick, segmented, and a mucusy-white, yellow-tinged on the surface. *Ancient,* and filled with a living force, a life rooted deep. As if the rootlets had reached far deeper than their length, as if they had pulled some ancient power from deeper in the Earth than humans have ever gone. As if mountains, immeasurably old, had grown under the soil. As if their power had found their way into these rootlets, into this plant.

As you look at the roots you feel something odd happen to your brain. Some older part, some ancient, reptilian part, is being stirred into life. You feel a movement deep in your brain. Some lizard-scaled being, huge and dark, lidless eyes staring, turns over. Shifts its weight. Then it blinks at you.

And then on the field of your vision you see the plant immersed in wetland bog. Its exterior boundaries lose definition, pale, thin, disappear. And you see a flow of bad air moving from the soil up through the plant, up and out, into the air of the world. As if the plant were some living channel through which the Earth were breathing, ridding itself of some stagnant thing.

Then the view changes, and standing at some higher point, you see the salmon coming back from their long journey to the sea, finding their way again to the stream of life in which they were born. And these plants are watching, hooded sentinels standing along banks of ancient streams, welcoming them home.

You notice then that you have been sitting, looking at the plant, for a long time, without noticing the movement of time at all. It is as if time itself had stopped. And you notice that your breathing is exceptionally calm. Not deep. Not shallow. But easy, whole. The lungs seem to be not separate things, but only an extension of the atmosphere, easily filling and emptying. Every part of them is taking in life from the world in which they are embedded. They seem to rest, to be nestled, in atmosphere.

As you awaken further from the trance into which you have fallen, you notice the colors around you once more. They are more alive. Somehow enhanced. Deeper. Luminescent.

Sounds come next. The muted sounds of life in ancient wetland forest. They shimmer with life, as if life is flowing along streams of sound, entering you, touching you. And you are caught up in them. Your whole body feels alive now, thrilling to the touch of light and sound.

After that come the smells. Your lungs breathe deeply, taking them in. And as they flow within you, they find purchase in the deepest parts of you. Almost as if they were a food. You can feel them enter into you as if your very cells were eating the smells, as if they were taking life from them.

Your whole being rests easily in this *place* in which you find yourself. It seems right, and you wonder how you ever forgot it, ever lost touch with this kind of beingness, this way of living. You look once again at the plant and *feel* the power of its medicine, the medicine that has just entered into you and changed you.

Then you sigh, breaking state, shake yourself, and breathe normally. You take the clippers now and cut the leaves from the root. Place them in the hole, the seed-laden flower stalk underneath. And feeling a slight regret at the beauty left behind (for the luminescent leaves will soon turn brown and lose their gleaming aliveness), you take the shovel, the clippers, and the pack they came in, and make your way out of the forest.

You place the plant on your herb table and let it dry, checking it each day. It takes awhile to dry completely; the roots are flexible, pliable, *mucilaginous*. But finally they dry, and you take some of the root and grind it into a fine powder.

As you lift the lid off the grinder, a fine mist of powder comes up with it. A puff of light, white smoke, eddying over the edge of the grinder, snaking its way under the edges of the lid that you hold in your hand.

You bend over slightly, putting your nose into the smoke,

this plant cloud

and you breathe in. Minutes later, you realize you haven't moved. You are held, suspended in time. A great calmness has spread through your body, and you feel *integrated* with the world. Your breathing is deep and calm and very, very easy.

You take the powdered root and put it in some alcohol and water, making a tincture. You visit it each day, talking to it, sending your caring into it, shaking the bottle to mix it well. In a few weeks it is done, and you are ready to take the lid off again. You can see the tincture through the clear glass of the bottle. It is a translucent living tan

sort of

with hints of gold. You pour off the liquid. Press the mushy, wet mass to extract the rest. And put it all in a brown bottle, out of the sun.

As a dropper is filled, a long, thin line of the tincture pulls up with it, then lets go, like a liquid rubber band, softly snapping and dropping back into the bottle. The taste is slightly sweet, earthy, *airy*. The tincture is mucilaginous, softly coating to the tongue. It moves through the mucous membranes of the tongue and then the breath comes deeper; a wild, powerful joy surges through the body. It seems as if you can run for miles, without being short of breath, without your lungs puffing-bellows at the end.

Then you put a label on the bottle, and write "skunk cabbage" on it, and know that this captures none of the living reality of the plant. But, inside you, the *name* of the plant resides, and you can call it up in memory, say it anytime . . . though not in words.

SECTION ONE

VERIDITAS

The natural world is a spiritual house, where the pillars, that
*   are alive,*
let slip at times some strangely garbled words;
Man walks there through forests of physical things that are
*   also spiritual things,*
that watch him with affectionate looks.

As the echoes of great bells coming from a long way off
become entangled in a deep and profound association,
a merging as huge as night, or as huge as clear light,
odors and colors and sounds all mean—each other.

Perfumes exist that are cool as the flesh of infants,
fragile as oboes, green as open fields,
and others exist also, corrupt, dense, and triumphant,

having the suggestion of infinite things,
such as musk and amber, myrrh and incense,
that describe the voyages of body and soul.

— CHARLES BAUDELAIRE

CHAPTER SEVEN

THE DOOR INTO NATURE

Seeking for truth I considered within myself that if there were no teachers of medicine in this world, how would I set to learn the art? Not otherwise than in the great book of nature, written with the finger of God. I am accused and denounced for not having entered in at the right door of the art. But which is the right one? Galen, Avicenna, Mesue, Rhais, or honest nature? Through this last door I entered, and the light of nature, and no apothecary's lamp directed me on my way.
— PARACELSUS

Oh, I understand! You were being a bridge.
Well, that's nice, bridges are important.
But you know, the only problem with being a bridge
is that you, yourself,
never get to cross over.

— NAN DEGROVE

Be not discouraged, keep on, there are divine things well envelop'd. I swear to you there are divine things more beautiful than words can tell.
— WALT WHITMAN

Yₒᵤ MUST ASK YOURSELF, IN THE BEGINNING, if you truly want to communicate with plants, just what is the status of the plant? Just *how* do you really feel about it? Is that plant there, the one near to your hand, your equal? If you do not feel that it is at least the same to you as a human being (it is better if you understand it is superior), then I am not sure it will talk to you.

Let us assume,

just for the sake of argument

that women are not the equal of men.

because they do not think as well

Now, go and talk to one and see how well you do.

Some of us have been raised well and some have not, but all of us have learned to be impolite to plants. (How much intelligence does it take anyway, as Larry Niven said, to sneak up on a carrot?) It is no wonder that they talk so seldom now

what about psilocybin?

or that the only ones we can hear are psychotropic, or insistently invasive, like kudzu. (The *loud* ones!) The cultivation of a delicacy of perception allows us to hear the quiet ones, the ones whose doctoring is most subtle. The polite ones. The ones who wait for us to speak before they respond.

The first step in learning to talk to plants is cultivating politeness, realizing that the pine trees that have been here for 700 million years must have been doing something before we came on the scene a mere million years ago.

besides pining away for our existence

The first step is to respect our elders.

Pine trees know a great deal more than we ever will about being pine trees and about what pine trees do. So all that nonsense you learned in school has got to go, especially the botany. In ordering plants (around), Linnaeus's voice has become so LOUD that every other sound is drowned out.

but I like saying Pinus

You are learning a different kind of language now, and you must be suspicious of the word. Words are the domain of the linear mind; only the heart can hear the language of plants. And words kill the perceptions of the heart.

> *How difficult it is not to put the sign in place of the thing; how difficult to keep the being always livingly before one and not to slay it with the word.*
>
> — GOETHE

This unlearning is difficult to do. It takes years. You will trail the scattered pages of dead learning behind you for decades as you go. The first step is the simple/hard one, depending on your orientation. Simple if you have a predisposition for it, hard if you have no idea what I am talking about. (Neocortex freaks will tell you this is all nonsense.)

The goal now is not to make plants manageable, but to make them visible. And only outlaws can see plants truly.

"Nature is," as Henry David Thoreau understood so well, "a prairie for outlaws." Those who go into Nature become, of necessity, *uncivilized*. Thoreau was well read. He knew that the word "civilized" comes from the Latin *civilis*, meaning "under law, orderly."

ah, his little joke

Civilis itself comes from an older Latin word, *civis*, meaning "someone who lives in a city, a citizen." Those who go into wilderness, into Nature that has not been tamed, are no longer under (arbitrary) human law, but under the all-encompassing, inevitable law of Nature. They go out from under human law. They are no longer citizens, they are not orderly, they are not civilized—they are outlaws. When you go into wilderness, something happens, something that civilization does not like. (That's why they cut it down, you know.)

> *I am afraid of the cities. But you mustn't leave them. If you go too far you come up against the vegetation belt. Vegetation has crawled for miles towards the cities. It is waiting. Once the city is dead, the vegetation will cover it, will climb over the stones, grip them, search them, make them burst with its long black pincers; it will blind the holes and let its green paws hang over everything.*
>
> — JEAN-PAUL SARTRE

Ordered regularity disappears in the wilderness, and people who live too long in wilderness also lose their regularity, their orderedness.

their willingness to be ordered

The discomfort with disordered nonlinearity, the fearful desire to be so *regular* and clean-shaven, has an antidote: the green, living intelligence of plants.

Veriditas

You must not mind getting their green paws on you. Only if you walk into Nature will you discover they are not paws at all but something else entirely. But this means immersing yourself. There is no place *within* Nature for observers, for conservative thinkers. The door that is sought never opens for the reductionist.

> *Those who hope to be reasonable about it fail.*
> *The arrogance of reason has separated us from that love.*
> *With the word "reason" you already feel miles away.*
>
> — KABIR

Because we are taught so many untruths about what we can know, about what Nature is and is not, the first step in gathering knowledge from the heart of the world is to go *into* the world on your own, abandoning your preconceptions.

the first act of courage

No expert can tell you what is there. No book knows the living reality of it. (For no book, not even this one, is a real, living thing.)

> *Everyone has to seek nature for himself.*
>
> — MASANOBU FUKUOKA

It is a living experience that you will find, not a mental construct. The things you think you know, that you have been taught, will get in your way if you do not agree, at least for now, to abandon them.

> *It is only when we forget all our learning that we begin to know. I do not get nearer by a hair's breadth to any natural object so long as I*

presume that I have an introduction to it from some learned man. To conceive of it with a total apprehension I must for the thousandth time approach it as something totally strange. If you would make acquaintance with the ferns you must forget your botany. You must get rid of what is commonly called knowledge of them. Not a single scientific term or distinction is the least to the purpose, for you would fain perceive something, and you must approach the object totally unprejudiced. You must be aware that nothing is what you have taken it to be. . . You have got to be in a different state from common. Your greatest success will be simply to perceive things as they are.

— HENRY DAVID THOREAU

How delightfully difficult it is to accept the reality of our ignorance. The door is *in* Nature, but only by giving up what you think you know about Nature, by being willing to know nothing, is the door found.

We should abandon all our preconceptions, most of which are afterward found to be absolutely groundless and contrary to facts. The final appeal must be made to the plant itself and no evidence should be accepted unless it bears the plant's own signature.

— JAGADIS CHANDRA BOSE

CHAPTER EIGHT

THE NECESSITY FOR ACUITY OF PERCEPTION

Nature will bear the closest inspection. She invites us to lay our eye level with her smallest leaf, and take an insect view of its plain.

— HENRY DAVID THOREAU

Thoreau was capable of true patience in observing the nonhuman world, and he exclaims in one passage, "Would it not be a luxury to stand up to one's chin in some retired swamp for a whole summer's day?" If we've read Thoreau, we know that he would be perfectly capable of it.

— ROBERT BLY

The observation of nature requires a certain purity of mind that cannot be disturbed or preoccupied by anything. The beetle on the flower does not escape the child; he has devoted all his senses to a single simple interest; and it never strikes him that at the same moment something remarkable may be going on in the formation of the clouds to distract his glances in that direction.

— GOETHE

SENSING THE WORLD AROUND US IS THE NEXT ESSENTIAL STEP, for the linear mind halts its activity in the face of sensory input from the wildness of the world.

sensory input
takes the place
of internal chatter

Our senses are meant to perceive the world. They developed with and from the world, not in isolation. Using them is the act that opens the door that is in Nature.

There is more to seeing than meets the eye.
— NORWOOD RUSSEL HANSON

Evolutionists have long puzzled over the emergence of the eye. The slow, step-by-step development of this organ of perception is a wonder to them, unexplainable in Darwin's world. They continually fail to see that organs sensitive to light existed long before Earth modified them to be human eyes. They have been present in photosynthesizing plants for hundreds of millions of years.

and their emergence was not slow at all

Our bodies are not so different from those of plants (in spite of what you have heard). The ecological expression of the animal from the bacterial, with wisdom gained through plant metamorphosis, carried photosensitive cells into new forms of expression.

and all for a reason

These cells, like those still embedded within plants, shaped themselves in response *to* and in interaction *with* the objects of their affection.

[The eye] owes its existence to the light. Out of indifferent animal organs the light produces an organ corresponding to itself; and so the eye is formed by this light so that the inner light may meet the outer. . . . If the eye were not sunlike how could we perceive the light.
— GOETHE

Our sensory organs are meant to perceive the world. The sensory capacities of human ears were shaped by sounds of the world, our smell formed through long association with the delicate chemistries of plants, our touch by the nonlinear, multidimensional surfaces of Earth, our sight by the images that constantly flow into our eyes. Human senses emerged from immersion within the world. They are part of Earth, an expression of communicative contact subtly refined and shaped through long association. They have their identity *in* the world, and are meant to perceive the every-minute-of-every-day influx of sensory communications that flow into and through them.

Focusing on the continuous flow of sensory data coming from the world around us activates our sensing bodies as organs of perception, leaves the computer behind, and embeds us once again within the world in which our species was born.

So, allow yourself to sense once again. Allow your sensory perceptions to *be* your thinking. *Sense* instead of think. This is what the senses are meant to do.

it's time to come to your senses

Our senses are living organs intended to receive communications. They connect us to, interweave us with, the stream of informational energy that comes to us every moment of every day that we live. Focusing perception through the senses immerses the self in the Earth's sensory flows.

kind of like taking a bath in colors, sounds, and tastes

To immerse the self in the wash of the world's communications, to feel the touch of the Earth through the body, brings the entire body alive. The way it was when you were young.

> *In youth, before I lost any of my senses, I can remember that I was all alive, and inhabited my body with inexpressible satisfaction; both its weariness and its refreshment were sweet to me. This earth was the most glorious instrument, and I was audience to its strains. To have such sweet impressions made on us, such ecstasies begotten of the breezes! I can remember how I was astonished.*
> — HENRY DAVID THOREAU

Attending to sensory communications from the world dissolves the boundary between self and world. It is a crucial act in reconnecting ourselves with the life of Earth. Sensory perception is the natural and right blending of inner and outer.

The *linear* mind is what creates the boundary *line* between us and the world. Location of consciousness in the brain closes the door to Nature. But the door is unlocked.

> *We do not need to please the doorkeeper, the door in front of us is ours, intended for us, and the doorkeeper obeys when spoken to.*
> — ROBERT BLY

Perceiving through the senses opens the door. The more sensitivity we cultivate to sensory flows, the more directly we perceive with our senses, and the wider the door opens.

> *It was a pleasure and a privilege to walk with [Thoreau]. He knew the country like a fox or a bird, and passed through it as freely by paths of his own. He knew every track in the snow or on the ground, and what creature had taken this path before him. Once must submit abjectly to such a guide, and the reward was great. Under his arm he carried an old music-book to press plants; in his pocket, his diary and pencil, a spy-glass for birds, microscope, jack-knife, and twine. He wore a straw hat, stout shoes, strong gray trousers, to brave shrub-oaks and smilax, and to climb a tree for a squirrel's nest. He waded into the pool for the waterplants, and his strong legs were no insignificant part of his armor. On the day I speak of he looked for the* Menyanthes, *detected it across a wide pool, and, on examination of the florets, decided that it had been in flower for five days. He drew out of his breast-pocket his diary, and read the names of all the plants that should bloom on this day, whereof he kept account as a banker when his notes fall due. The* Cypripedium *not due till tomorrow. He thought that, if waked up from a trance, in this swamp, he could tell by the plants what time of the year it was within two days. . . His power of observation seemed to indicate additional senses. He saw as with microscope, heard as with ear trumpet, and his memory was a photographic register of all he saw and heard.*
> — RALPH WALDO EMERSON

Habitual location of consciousness in the brain—our long immersion in the analytical mind—atrophies the capacity for sense perception. What's worse, we have been taught that the senses are unreliable as organs of perception. Linear fanatics' biased reporting has terrified us into distrust of what our senses tell us, terrified us as to their reliability.

People no longer tread over the bare earth. Their hands have drawn away from the grasses and flowers, they do not gaze up to the heavens, their ears are deaf to the songs of birds, their noses are rendered insensitive by exhaust fumes, and their tongues have forgotten the simple tastes of nature. All five senses have grown isolated from nature.

— MASANOBU FUKUOKA

Thus, the second act of courage is deciding to trust your senses. To use them to perceive the world around you. To use them as they are meant to be used, as a channel to the world in which you were born, from which you have been expressed, and that is communicating to you through your senses every moment of every day.

The careful observer can glimpse the seemingly impossible even with the unaided eye, a fact which forces one to prostrate oneself in adoration before the mysterious origin of all things.

— GOETHE

And to use your senses most productively, you must go out from the cities. You must leave the Euclidean geometry of skyscrapers and rectangular living rooms, move away from all those Cartesianly coordinated streets. You must find a place where Nature is not buried under concrete and asphalt.

where it is out from under the law

To begin to cultivate depth perception of Nature, to gather knowledge directly from plants, go to the plants themselves. Take a walk someplace wild, where the civilized do not go.

Go to the pine
if you want to learn about the pine,
or to the bamboo
if you want to learn about the bamboo.

— BASHO

Take a few deep breaths when you arrive, settling yourself down deep in your body. Then begin to walk. And as you walk, become sensitive to the feel of Earth under your feet. Notice how it forces your body to move differently than a sidewalk does, how each tiny perturbation in the non-Euclidean reality of Nature forces upon you a multidimensionality of movement.

Different muscles are needed in Nature than those that are used in cities.

Now, let yourself drop down into your feet, your feet becoming sensing organs themselves, supportive organs. Let go of holding yourself up, let your feet hold you. And let the reality of Earth come through their touching into you. Feel the way that is before you.

When you get used to this, let the Earth hold your feet.

a mother's embrace

As you continue to walk, become aware of the sounds that surround you, that are continually touching you through the delicate surfaces of your eardrums. Become aware of those oh-so-delicate vibrations. The tiny movements and flutterings of life.

> *The earth-song of the cricket! Before Christianity was, it is. . . Only in their saner moments do men hear the crickets.*
> — HENRY DAVID THOREAU

The quiet susurrus of the wind across the green fingers of the grass. The flutter of a bird's wings. The tiny sounds that only children hear. Focus on them, let them grow in your awareness until they are all that you hear.

> *The catbird, or the jay, is sure of the whole of your ear now. Each noise is like a stain on pure glass.*
> — HENRY DAVID THOREAU

Buddhists have long worked to teach people to still the chattering mind, understanding that this is an essential step. Some of those techniques, when inserted into the illusionary dualism separating spirit and matter, all too often create an antagonistic relationship with a part of us that is, and is meant to be, an ally. You cannot stop the linear mind, leaving nothing in its place.

it resents destructive approaches

The work is simple. Do something else instead.

You must walk so gently as to hear the finest sounds, the faculties being in repose. Your mind must not perspire.

— HENRY DAVID THOREAU

As your body becomes more and more alive through the activation of your senses, *sensing* is what you do instead of thinking. Sensing takes the place of thinking. Awareness is focused through your senses, attentively noticing all that you sense. It has no time for thinking now. Your consciousness begins to move out of the brain, leaving the analytical mind behind. You begin to find the world that our ancient ancestors knew so well.

> *The plant hunter learns to sharpen her senses.*
> *On the lookout: an alert. Are plant walkers about?*
> *Soul-stealing plants? She tastes*
> *bitter healers, smells sweet-leafed herbs,*
> *finds fiber plants and grain plants,*
>
> *cold mornings,*
> *mammoth-skin huts.*

— DALE PENDELL

As you continue to walk, let your eyes notice the colors around you. Focus on your visual sensing of the world now. The green of plants exists in a thousand different shades. Let your sense of sight flow out from you and touch those colors, noticing the delicate interplay of shadow and light, the minute shadings of color from one plant to the next, from one leaf to the next.

Always, if you allow yourself to notice, one plant will seem more interesting than all the others. It is to this one that you must go.

Focus on this plant that has called you. Let your feet take you to it. Sit before this plant as it rests nodding in the sun.

The spiritual master . . . bows down to the beginning student.

— KABIR

Let your eye focus on this plant's leaves. Notice their shape, their orientation in space. How they are arranged along the stem. Notice the shape of the stem, its color, and the color of the leaves. There is a specific texture to the leaves, a quality of smoothness or roughness. Let your eye sink into their depths and see them in their tiniest instant of time.

Keep your botany out of this! Do not classify! Do not use big, scientific words!

The ones who see true nature are infants. They see without thinking, straight and clear. If even the names of plants are known, a mandarin orange tree of the citrus family, a pine of the pine family, nature is not seen in its true form.
— MASANOBU FUKUOKA

As you describe, speak as if you were a four-year-old. "Fuzzy," a child says. "Pointy."

So much of man as there is in your mind, there will be in your eye.
— HENRY DAVID THOREAU

Now, touch the leaf, *feel* it with your fingers, those sensitive, sensory extensions of your self. Let the sensation of touch fill you. Immerse yourself in the living texture of leaf until it is all that you know.

Forget everything, for example, but the leaves of plants and trees. Notice those in your garden or a park or along streets, or in the country. No two just alike! So different in shape and form and thickness and texture and length and position on the tree or the plant or the twig or the stalk as hardly to be the same sort of thing.
— LUTHER BURBANK

Bend closer to the plant, so its leaf is near your nose. Rub it lightly across your skin, feel how that feels. Now smell. Take in a long, slow breath as if you were taking in the subtle, deep scent of someone you love. Let your awareness focus down until this smell is all there is in your awareness.

Immerse yourself in this smell. Savor it. Let the nuances of the scores of delicate chemical compounds now touching the sensory receptors in your nose enter you. There are delicate shadings. Smells that possess the same layering as the colors of the leaves. Smells that have much to communicate.

I have always been sensitive to odors, so that I could detect them, pleasant or disagreeable, when they were so slight that no one about me was conscious of them. My sense of touch is almost as acute as that of Helen Keller, who visited me just a short time ago.
— LUTHER BURBANK

Now hold the living leaf in your mouth. Become aware of how it feels to your tongue. Sit a moment, let yourself deepen into the experience. Don't worry if you feel foolish. Breathe slowly into this experience. How does your body respond? Does it like this plant or not? Now sit back, slowly letting the leaf go.

Then take a small piece of the leaf and eat it.

Don't put that in your mouth!

How does it taste? What is its flavor? Bitter? Drying? Green?

Have you become so used to the taste of domesticated plants that the taste-reality of this plant is unpleasant? Let your preconceptions about taste go and notice how your body responds to this taste. How you respond to it.

What if it's poisonous?

One of our greatest fears is to eat the wildness of the world.

Our mothers intuitively understood something essential: the green is poisonous to civilization. If we eat the wild, it begins to work inside us, altering us, changing us. Soon, if we eat too much, we will no longer fit the suit that has been made for us. Our hair will begin to grow long and ragged. Our gait and how we hold our body will change. A wild light begins to gleam in our eyes. Our words start to sound strange, nonlinear, emotional. Unpractical. Poetic.

> *Children are attracted by the beauty of butterflies, but their parents and legislators deem it an idle pursuit. The parents remind me of the devil, but the children of God. Though God may have pronounced his work good, we ask, "Is it not poisonous?"*
> — HENRY DAVID THOREAU

Once we have tasted this wildness, we begin to hunger for a food long denied us, and the more we eat of it the more we will awaken.

> *Part of us still knows we need the Wild Redeemer.*
> — DALE PENDELL

It is no wonder that we are taught to close off our senses to Nature. Through these channels, the green paws of Nature enter into us, climb over us, search within us, find all our hiding places, burst us open, and blind the intellectual eye with hanging tendrils of green.

The terror is an illusion, of course. For most of our million years on this planet human beings have daily eaten the wild. It's just that the linear mind knows what will happen if you eat it now.

But we've gone astray with this, distracted from our task.

funny how fear does that

Still, it's a good reminder. When your hair begins to grow long and you think strange thoughts, sometimes you will wonder what is happening and will become afraid.

all of us do

In Nature, human markers fade, lose significance. It takes awhile to learn the old markers again, to see the path that ancient humans took before us. In kindness, learn how to comfort yourself, to hold yourself as you would a child that is afraid of the light. (I suppose you could learn the poisonous plants first if you need to; there aren't very many.) For on this journey, you mostly have yourself for company.

in the beginning

It helps if you become your own best friend
and find out what is true about all this for yourself.
Open the door and take a look around outside.
The air is shining there,
and there are wonders,
more wonderful than words can tell.

This morning I am washing citrus storage boxes by the river. As I stoop on a flat rock, my hands feel the chill of the autumn river. The red leaves of the sumacs along the river bank stand out against the clear blue autumn sky. I am struck with wonder by the unexpected splendor of the branches against the sky.

Within this casual scene the entire world of experience is present. In the flowing water, the flow of time, the left bank and right bank, the sunshine and shadows, the red leaves and blue sky—all appear within the sacred, silent book of nature.

— MASANOBU FUKUOKA

CHAPTER NINE

FEELING WITH THE HEART

We all walk in mysteries. We do not know what is stirring in the atmosphere that surrounds us, nor how it is connected with our own spirit. So much is certain—that at times we can put out the feelers of our soul beyond its bodily limits; and a presentiment, an actual insight . . . is accorded to it.

— GOETHE

A man has not seen a thing who has not felt it.

— HENRY DAVID THOREAU

There is no way of expressing that mountain which goes beyond a mountain. Nature can only be understood with a nondiscriminating heart.

— MASANOBU FUKUOKA

And now here is my secret, a very simple secret; it is only with the heart that one can see rightly; what is essential is invisible to the eye.

— ANTOINE DE SAINT-EXUPERY

USING DIRECT PERCEPTION TO LEARN the medicinal powers of plants is not a spectator sport.

Every true vegetalista has to meet Sacha Runa, the forest person, face to face. In the Jungle.

— DALE PENDELL

Eventually you have to move from looking and go into feeling, realizing that feeling is a sense, too. Not the touch of the fingers, but the touch of the heart. This kind of touch has another dimension, deeper than that possessed by the fingers.

and as we touch, so are we touched

Everything has a hidden face. Hidden, not in the sense that it is intentionally concealed, but in that it can only be seen with different eyes than the physical. A different mode of perception must be used. The hidden face of Nature can only be seen with the heart.

As you sit with the plant, focusing on its sensory attributes, begin to

s l o w   d o w n   n o w.

Become aware of the feelings that arise in you as you sit by the plant. How do you feel? Now you are learning to see—not merely the physical form of things, but the meanings that each thing expresses.

Everything we encounter in the wildness of the world gives off its own electromagnetic pulse of communication. These waveforms are filled with meanings, living communications that touch us and that we experience as feelings.

At dessert, Goethe had a laurel, in full flower, and a Japanese plant, placed before us on the table. I remarked what different feelings were excited by the two plants—that the sight of the laurel produced a cheerful, light, mild, and tranquil mood; but that of the Japanese plant, one of barbaric melancholy.

— JOHANN PETER ECKERMANN

Because we have been taught for so long to disregard these kinds of feelings, it may be hard to let yourself notice them. Begin by allowing yourself to describe these plant-generated feelings in any way they come to you. Take the step of letting them come into consciousness and emerge into words. Do not control the words or make them big and analytical. Let them emerge of themselves, in their own form. Give yourself permission to say out loud what they are, no matter how foolish they might seem to your linear mind.

It is a rare qualification to be able. . . to conceive and suffer the truth to pass through us living and intact.

— HENRY DAVID THOREAU

Because of our long habituation in the linear mind and the things we have been taught about the livingness of the world, this is the hardest thing of all—to give reality to the feelings that flow into us from the world itself.

A two thousand-year-old tree
in an ecosystem filled
with a
> *tumultuous,*
>> *complex,*
>>> *riot*
>> *of interacting plant species*
> *feels*
markedly different
than a lone sapling
surrounded by grass,
stark in the front yard
of a new housing development,
or the Norfolk pine
leaning drunkenly
in the corner of the kitchen.

The green,
> *orderly*
> *lawns*
surrounding children's homes
do not bear any relationship
> *to the up-and-down,*
>> *uneven landscapes*
>>> *filled with giant,*
>> *craggy outcroppings*

of the immeasurably ancient stones of Earth
that wild landscapes often possess.

A calm pond lends us serenity,
yet when its waters
are disturbed
by wind
are we not also disturbed?
our emotions unsettled?

Where is it
that our feelings
really come from?

Giving reality to the feelings that come to us from the world directly contradicts the Western insistence on the linear mind and the (presumed) unreality of the living soulfulness of the world around us. Doing so breaks a cultural agreement that is powerful and strong and deeply embedded within us.

We have lost the response of the heart to what is presented to the senses.
— JAMES HILLMAN

Embracing the reality of the feelings that come to us from the world is the first step in the decolonization of the soul. In this moment, the linear mind is truly left behind. This is the moment you begin to use a different mode of cognition—the moment you begin to think with your heart.

the third step

Most of us have been taught that feelings only come from within us. For those of us who wish to learn directly from the wildness of the world, to learn directly from plants the medicinal uses they possess, it is essential to begin to feel with the heart. To do so you must go to meet the plant with your most vulnerable self. You must open your heart and let the plant's living communications flow into you, weave through you. You must receive what it has to offer.

Leave the well-worn road you have traveled so long. Have the courage to speak the intangible.

to write it down in a journal

Write out all the feelings you notice now. Let yourself say them without trying to make them pretty, grown up, or elegant.

There will be one or more primary feelings: mad, sad, glad, or scared. Then a number of secondary feelings: a unique blending of the primary feelings into more subtle forms, like the million blendings of colors from an artist's palette. These secondary feelings are encodes of more complex communications from the plant. Just as plants create primary and secondary chemistries, they create primary and secondary electromagnetic pulses, primary and secondary feelings.

As it feels the impacts of these feeling complexes, your body will respond at a level deeper than your conscious awareness. There will be an immediate physical articulation in response to what you are sensing.

Your body's response
can be exceptionally subtle

Because it is out of your conscious awareness, you must notice everything that your body does during this process, everything you feel, every stray thought that comes into your head, no matter how insignificant, unrelated, or ridiculous it seems.

You are learning a new language now. In this process, your body is your best friend and most important teacher. You must learn to honor it once more, to not denigrate or distrust it as you have been taught in school. It knows and will teach you. If you let it. If you respect it.

It makes a wonderful difference whether we find in the body an ally
or an adversary.

— GOETHE

So pay attention to everything that your body does as you sit with the plant, everything you think and everything you feel. Cultivate a perceptive awareness of all these responses, learn to be sensitive to the least movements of your self in all the forms this may take. Write it all down.

And give up your preconceptions. For if you have an assumption about the form in which the knowledge will appear, you will overlook much that is important.

The young woman beckoned to me, upset.
"What is it?" I asked.
She took my arm, led me to a sheltered spot.

"I tried the exercise," she said, taking a deep breath, "and noth-ing happened." She brought her hands to her chest, took another deep breath, seemed about to cry.

"Really?" I asked.

"Yes, I tried it," she said, placing her hands on her chest again, taking another deep breath. "Nothing happened."

"How do you feel?"

"Sad," she said.

"What plant did you sit with?" I asked her.

"That one over there. Mullein," she said, pointing it out where it stood, its high stalk gently moving in the breeze.

"But don't you know that mullein is used for the lungs, for help-ing breathing to be more easy? Deeper?"

"No," she said, looking confused.

"And people hold a lot of sadness in their lungs. They close them down, compress them, so they won't feel the sadness sometimes."

I took her hands gently, said, "It is no accident that you are breathing more deeply, that you are about to cry. Every time you speak of the plant you breathe deeply and touch your chest, your lungs and heart. You must learn to pay attention to everything that happens. The plant–human communication is always language. It is not always words."

All phenomena, when focused upon, will generate an intimation of a particular mood or quality within you. We are daily touched by the world within which we are embedded, we feel that touch upon us in the thousands of nameless feelings we experience each day. They flit over the surface of our consciousness like shadows across a grassy meadow. In paying attention to them, they come forward into consciousness and begin to reveal their secrets, for each emotion registers the impact of a particular meaning that has touched us. They are *transforms* of information, of communications, from the world around us. These transforms contain extremely condensed and elegant communications about what we are encountering.

> *Though the gods have the power of speech*
> *more often they choose a flower or plant:*
> *elder leaves pressed on a blotter,*
> *or spring buds emerging from a winter stem*

These messages they send—
so ordinary we usually miss them:
an easy laughter and lightness,
or legs casually crossed and touching

The way a serpentine dike blends seamlessly into bedrock
or the way two possible lovers move,
starting and stopping, passing and pausing,
on an April trail

The subtlest oracles are always the most obvious—
seeing what is clearly in front of us the most difficult:
a butterfly hatching from a ruptured dream,
or a splintered tree rooting in the soil where it fell

— DALE PENDELL

Paying attention to the feelings of things and writing them down is a good beginning. It starts to train you in a specific skill. It is like learning to ride a bicycle.

or a unicycle

It takes a lot of practice to gain facility with it, to find the balance point, to trust yourself to it. After all, the experts have been telling you your whole life that there is nothing out there, no communications, no intelligence, no meaning, no sacredness or soul. Still, our ancestors walked this way before us. We are meant to feel the touch of the world upon us.

The true was already found long ago.

— GOETHE

This initial intimation, the impression or mood, the feeling of the plant, is the beginning of your connection to its being. Firmly anchor it into your experience.

don't forget it

This initial intimation is the key to unlocking the mysteries of the plant itself, the way to understanding its uses as medicine. It is to this initial impression that you will return over and over as you refine your knowledge of the plant you are studying.

Keep practicing; it is repetition, repetition, repetition that will habituate the skill. The more you do this, the better at it you will become. The more plants you experience, the better you will find the process to be.

Know that these feelings are encoded communications from the world around you, transforms of messages. But they are not feelings from which you can remain distant. To feel them is to connect with the world around you, to allow your life to interweave with that of the plant and the world in which the plant lives. It is the beginning of an intimacy with life, a mode of living in which you are never alone, in which communications come from the world to you and go out from you to the world. It is a way of being.

At that moment when each thing, each event presents itself again as a psychic reality, then I am held in an enduring intimate conversation with matter.

— JAMES HILLMAN

You can extend this, deepen it, go even further. Feeling the touch of the plant upon you is only the first step. You can also touch it in return, intend communication with it as well.

While sitting with the plant, keeping strongly in mind its living reality, become aware of your heart. Breathe in through your heart, breathe in the feelings of the plant that are coming to you.

let them deepen, become stronger

Now, feel the nonphysical energy field of your heart emanating out from you. Envelop the plant with the field that your heart is creating with each beat. Feel it holding the plant within it.

Let it touch the plant the way your eyes touched the thousand different shades of green on the leaves. Let it delicately touch the near infinite shades of meaning/feeling that the plant is giving off. Let your human-touching and the plant-touching interweave, blend together. Feel your heart touch the plant, and in that touching, connect with it at all possible points of interface.

Once you see it, you know it
was there all the time, so why
is it all such
*   a big deal? And why*
do we keep forgetting?

— DALE PENDELL

Now, let the beauty of the plant affect you. Notice how much you care for it. Send out from your heart the love that you feel. Encoded in the complex, multivaried field of your heart are the feelings of caring you are now generating. And the plant, like all life, will take them in, respond to them, altering its communications in turn.

You will feel yourself slowing down as you do this, beginning to breathe more deeply as this progresses. This is the sign that you are moving more deeply into the heart as an organ of perception. Your entire physiological functioning is altering.

> *your eyes will become soft-focused*
> *your breathing slow and deep*

As you develop your sensitivity, you can feel the plant begin to move toward you, respond to you, engage with you, entrain with your heart. You can tell, when you pay close attention, the moment when the two of you have established rapport.

> *If the individual temporarily abandons human will and so allows himself to be guided by nature, nature responds by providing everything. To give a simple analogy, in transcendent natural farming the relationship between humanity and nature can be compared with a husband and wife joined in perfect marriage. The marriage is not bestowed, not received; the perfect pair comes into existence of itself.*
> — MASANOBU FUKUOKA

In that moment, send a request out from the deepest recesses of yourself. Ask the plant how you can use it as medicine. Tell it of your need.

> *Anything will give up its secrets if you love it enough.*
> — GEORGE WASHINGTON CARVER

There will be a response. Though you may have to pay attention to your body, your feelings, and the odd stray thoughts or pictures that pop into your mind to perceive it. Sometimes a phrase will, of itself, emerge into the mind.

> *for making rough things soft*

Or perhaps a picture will flash on the field of your interior vision.

Then I saw a tiny baby in a cradleboard. It was wrapped in soft leather but between the leather, and the baby was this plant, powdered, patted on, covering and embracing its skin.

Or you will breathe deeply. Or a flush of relaxation will flow through your body and your skin begin to tingle.

or perhaps all of these

You may want to go look up the medicinal actions of the plant that you have been sitting with.

to convince yourself that all this is real

To see that what you are receiving has some basis in reality, that it is in the books of the "experts" as well. Take it one step at a time, take as long as you need. It takes a long time to really trust this most ancient of skills,

to reclaim it as your own

for our colonization has been deep and long and all of us have forgotten much.

This process works best at first if you begin with plants that you instinctively feel drawn to. There is already something happening with those plants, some intimation of connection that touches you through this instinctive desire for closeness with these particular plants instead of others. The plants you feel most drawn to are those with which your heart already feels a kinship.

(There is another kind of plant that will draw you, too. The ones that will thrust themselves upon you in spite of your desire to notnotice them. The weed that will not go away and that you continually notice in irritation, the plant that regularly trips you as you walk through the field. Such plants are often some of the most powerful medicines you will find. They stir something in your unconscious, breaking through your habituated notnoticing, and intrude on you until you begin to take a real look at them.)

Then there are the one or two crazy ones
that will freak you out,
pick on you through your new sensitivity.
But coyote medicine . . . that's another story.
No need to be afraid.

> *It will be awhile before you run into them,*
> *only when some deep part of you is ready,*
> *and has cried out to them for help.*

Plants will, if genuinely asked, respond to you. They will teach you their medicine, as plants have always taught human beings. And though human beings may lose the knowledge of the medicinal uses of a plant, the plant always remembers what its medicine is. And they will tell you . . . if you ask. If you approach them with an open heart, open your senses and truly allow yourself to perceive them, they will always respond.

If you fail the first time, go again. For you may go to the sea as often as you wish.

> *You may go to the plant a thousand times;*
> *it will never turn you away.*
> *Just because someone once told you*
> *that you did something wrong*
> *doesn't mean that you did.*

You will eventually learn to hear. The plants don't mind if you practice, or if it takes some time, for they are the most caring of living beings. It's just that they like to be asked.

The problem with scientists—with those who think the world is a dead place—is that they never ask. They take . . . in the name of science.

> *We may force no explanation from [Nature], wrest no gift from her,*
> *if she does not give it freely. . . Nature becomes mute under torture.*
> *Its true answer to an honest question is: "Yea, yea; Nay, nay."*
> *Whatsoever is more than these cometh of evil.*
>
> — GOETHE

With plants, you must "talk to them like human beings," as the Winnebago Crashing Thunder's father said so long ago. "Then," he said, "most certainly will these plants do for you what you ask." This respect for your elders is essential, and the information flow that will come from them to you will contain all that you wish to know.

> *Taking without permission,*
> *the way that scientists do,*

is a form of rape.
The rape of Nature really is the rape of Nature.

You only have to love them, to feel the touch of their communications on your heart, and send out your deep request in turn. They will respond to you if you ask, for that is what plants are meant to do.

Writing down what you receive is not for memory's sake. It is only an elegant way to focus your mind on all the forms in which the communications appear. Once you do this, once you feel so deeply, you will remember the most important communications.

they will remember themselves

This deep intimacy and sharing flows directly from the heart into the memory centers of the brain; the hippocampus alters its functioning, and new neurons and neuronal pathways are created. The memories that are encoded are deep, the key to accessing them the remembered feelings of the event itself. When you recall the plant and this moment of touch, the memories will flood back as fresh as if they happened only moments ago.

It takes practice to refrain from inserting what you think you know into this process. You must remain with the thing itself. Allow it to speak to you in its own terms, hear it with the ears of a child, so that its true nature enters into you.

Clear and unprejudiced ears hear the sweetest and most soul-stirring melody in tinkling cowbells and the like (dogs baying at the moon), not to be referred to association, but intrinsic in the sound itself.
— HENRY DAVID THOREAU

The intentional expression of caring, attentiveness, and love alters the electromagnetic field of your heart, embedding new transforms of messages within it. This field, now carrying these new informational impulses, touches the plant field it is directed toward. This field takes in the information embedded within your heart's field; the living organism decodes the information and alters its functioning in response.

I then pass the whole day in the open air, and hold spiritual communion with the tendrils of the vine, which say good things to me and of which I could tell you wonders.
— GOETHE

The plants respond to the gesture of intimacy contained within the field projected by your heart. They respond, embedding new communications within their electromagnetic fields, which your heart takes in, decodes, and uses to alter its own functioning once again. You and the living phenomenon with which you are making contact entrain, and a living dialogue begins to flow back and forth, extremely rapidly.

this is reinhabiting your interbeing with the world

The flow of life to life and back again binds you into the web of life from which you have come and in which you belong. There is an ecstacy in the process, a coming alive again.

A thrumming of piano-strings beyond the garden and through the elms. At length the melody steals into my being. I know not when it began to occupy me. By some fortunate coincidence of thought or circumstance I am attuned to the universe, I am fitted to hear, my being moves in a sphere of melody, my fancy and imagination are excited to an inconceivable degree. This is no longer the dull earth on which I stood.

— HENRY DAVID THOREAU

We have an innate capacity to entrain ourselves, to establish a harmony of patterning, a rapport, with anything upon which our heart-attention is focused. When you emotionally hook yourself to a living thing, you anchor yourself to the nonlinear flow of its life. As your connection is deepened, you begin to flow with its life patterns; you absorb its meanings, its intelligence, and its particular point of view.

We must look carefully at a rice plant and listen to what it tells us. Knowing what it says, we are able to observe the feelings of the rice as we grow it. However, to "look at" or "scrutinize" rice does not mean to view rice as the object, to observe or think about rice. One should essentially put oneself in the place of the rice. In so doing, the self looking upon the rice plant vanishes. This is what it means to "see and not examine and in not examining to know." Those who have not the slightest idea what I mean by this need only devote themselves to their rice plants.

— MASANOBU FUKUOKA

Slowly, then ever more quickly, you will become aware of the living fields that are created by the life around you and the feelings they generate in you when you are open to them. You will intimately come to understand the truth that Pythagoras spoke so long ago: "Astonishing! All things are intelligent." You will understand, as the great Sufi teacher Hazrat Inayat Khan said, that "Everything is speaking in spite of its apparent silence." You will begin to communicate with plants, and they with you. Just as humans have always done.

> *He who sees into the secret inner life of the plant, into the stirring of its powers, and observes how the flower gradually unfolds itself, sees the matter with quite different eyes—he knows what he sees.*
> — GOETHE

Everything that you experience with the plant as you sit with it is important and bears some relation to its uses as medicine, its function in the ecosystem, its own life history and desires, and its relationship to humans and the world around it. Some of this may come powerfully in words, some only through a general sense of something that you may have more difficulty defining. You may feel a lightness of spirit when you are near the plant, a more positive mood. This ability to brighten the spirit, to lighten the load of the human predicament, to make it more bearable, is a medicine that many people need, and this quality should never be overlooked in a plant.

> *After all, when you take away the big words,*
> *this is all that antidepressants (are supposed to) do.*

This initial grouping of feelings, physiological responses, vague intimations, scattered linguistic descriptions, pictures that flash in the mind— are the beginnings of understanding the medicine. It is crucial to remember, however, that the feeling of this plant, the grouping of primary and secondary feelings that you experienced when you opened yourself to the plant, is the most important thing of all. These feelings are your connection to the living reality of the plant. The complex gestalt of plant communications that you experience as feelings is (not metaphorically, but literally) the medicine of the plant.

> *Everything in the realm of fact is already theory. . . Let us not seek something behind the phenomena—they themselves are the theory.*
> — GOETHE

The feelings *are* the medicine. At this moment, the medicine is simply encoded in one particular form. If you forget this, lose sight of it, lose the feeling of it, go off in your mind someplace else, you are losing touch with the most essential thing. This is the one true thing that you have come to the plant to experience. For every time you begin to gather information directly from the heart of the world, you must look for that

one true thing

that the phenomenon has to offer you. The one true thing is the complex of feelings that you experience from that phenomenon, that plant. It is the burst of communications that come from the thing being studied, the plant or landscape or ill person you are coming to know. It is not a thinking thing that you have here, this one true thing, it is a feeling thing. And this feeling is a unique living identity that must not be killed with the word.

Always a part of you will know when you are touched by the one true thing. We all know the true when we feel it.

This one true thing, this initial, powerful intimation, this burst of communication from the plant, is the most important thing of all. It is subtly altered as it passes into you, translated by your body and mind into feelings, senses, physiological responses, vague intimations, and scattered groupings of linguistic descriptions. This is the initial emergence of the medicine of the plant into a form usable to you as a healer. And this is fine in and of itself. It need never go deeper in order for you to be able to heal with this plant, to use it effectively to help people.

many doctors don't even have this

Over the years, as you use the plant for healing, your knowledge of it will deepen. Through long association, it will reveal more secrets to you. As long as you remain in touch with that one true, living thing, always treat the plant as a human being, and ask it for its help, your intimacy and dialogue will deepen.

Plants are like people; some you will not like, some you will be ambivalent about, some are boring, some are nice acquaintances that you will get to know slowly over the years, but some . . . some you fall in love with immediately and deeply and want to know intimately. You will want to know their life stories, know them as completely as you have ever known anyone.

There are processes for going further with this. To do so, you must bring fully into consciousness the meanings embedded in the moment of first contact.

The meanings that are embedded within that one true thing.

> *I have a feeling that my boat*
> *has struck, down there in the depths,*
> *against a great thing.*
>   *And nothing*
> *happens! Nothing. . . Silence. . . Waves. . .*
>
> *—Nothing happens? Or has everything happened,*
> *and are we standing now, quietly, in the new life?*

          — JUAN RAMON JIMINEZ

SECTION TWO

THE TASTE OF WILD WATER

It is good knowing that glasses
are to drink from;
the bad thing is not to know
what thirst is for.

— ANTONIO MACHADO

I remember gestures of infants
and they were gestures of giving me water.

— GABRIELA MISTRAL

CHAPTER TEN

GATHERING KNOWLEDGE FROM THE HEART OF THE WORLD

I wish it to be wholly understood what I have become to Nature and what Nature has become to me. If you wish to understand me only passably, you must know how Nature found me and I found Nature during our first encounter; then you will have the history and the exposition of my perceptions.

— GOETHE

Though you roam the woods all your days, you will never see by chance what he sees who goes on purpose to see it.

— HENRY DAVID THOREAU

Often I am permitted to return to a meadow
as if it were a given property of mind

that certain bounds hold against chaos,
that is a place of first permissions,
everlasting omen of what is.

— ROBERT DUNCAN

elegant.

For many people, this way of gathering plant knowledge remains only a vague sensitivity. But the perception of meaning and the elicitation of knowledge directly from phenomena, from plants, can be extremely elegant. The knowledge that is gained can be exceptionally detailed, more sophisticated than that found through (reductionist) science.

A great deal of information will come in the plant's initial gestural impulse, what the interior you interprets as linguistic groupings, flashes on your inner field of vision, physical responses, and complexes of primary and secondary feelings. This first impression bears much the same relationship to the plant as the intimations that come from a first meeting with an interesting person. To increase your knowledge your level of intimacy must increase. You must come to know the plant even better.

With plants, as with people, intimacy begins with a chance meeting. You may hear about a plant and its healing uses from another herbalist and find that you cannot stop thinking of it, see a picture of it in a book and find your attention captured, or encounter it by chance on a walk in the wild and notice that it seems special to you in some way. These chance encounters are the beginning of intimacy. The intimacy deepens when you take time to foster it, when you focus on the plant and take the time to really come to know it.

> *That aim in life is the highest which requires, as of a lifetime, to appreciate a single phenomenon! You must camp down beside it as for life, having reached your land of promise, and give yourself wholly to it.*
> — Henry David Thoreau

You can do this in many ways. One is to spend a great deal of time with the plants you are drawn to, to greet them with the seasons, see them in all manner of dress, come to know their moods and relations, and let your relationship grow with the years and close association.

This takes time, (as all worthwhile things do). You are establishing a deep intimacy with another living being. One that has a life as important to it as yours is to you. One that has a history, ancestors who have shaped the life you see now in this one moment of time. One that has hopes and dreams, purposeful existence. One that has other friends, offspring about whom it cares, troubles with which it must struggle each day.

> *A mother's love for her child and the plant's care for its seed possess an identity of essence. We are not separate from the world nor better*

than the other life forms that, like us, have been expressed out of Earth. The attributes and tendencies we possess are possessed by others and are only expressions of characteristics that are inherent in the world, common to all life forms. They are not unique to us.

As your relationship with the plant deepens through the years, as you come to know one another more intimately, knowledge will come of its own accord. One day you awake and, taking a walk, as you pass the plant in the field, of a sudden, deeper knowledge of the plant and its purposes, its uses as medicine, will flash into your mind.

When he abandons discriminating knowledge, non-discriminating knowledge of itself arises within him.
— MASANOBU FUKUOKA

This deeper knowledge becomes part of your understanding of the plant, part of the fabric of your relationship. The more years you are in active relationship, the more these bursts will come to you. The emergence of these bursts of knowledge depends a great deal on the closeness of your relationship with the plant. Not all plants will stimulate in you the same degree of intimacy.

some plants are just as boring as sociologists

Personal relationships with plants are like personal relationships with people. A whole range of experience exists, for there are many types of plants, each with its own personality, each of which will draw your interest and affection to differing degrees.

The slow emergence of deep knowledge over years of relationship and close association with a particular plant occurs in this fashion because the plant itself generates a certain kind of relationship. It is the kind of relationship that you and the plant are intended to have. It comes from your own particular natures.

With other plants, though, another approach is possible. With a plant like this, at the moment of first contact, you will feel a compulsion to know it deeply. You will feel drawn to it as if something outside of yourself were pulling the two of you together, as if you were destined to know one another. Like a new lover, you cannot stop thinking of it, feeling the feelings this new relationship engenders. You begin to feel that there is

Mentzryville, old glory. predigesta, ×
JUNIPER

some destiny in this meeting and that some truth—some newness of self—
is meant to come to you now. You are pulled into the plant, motivated to
see more deeply and with different vision. To see from another perspective, aslant to your normal orientation.

*Man cannot afford to be a naturalist, to look at Nature directly, but
only with the side of his eye. He must look through and beyond her.*
— HENRY DAVID THOREAU

With plants like these, a deepening of the process can occur, engendering a more rapid accumulation of knowledge and understanding. To make
this happen, you must take the moment of first contact, the memory of the
initial mood and feeling that came when the two of you met, and work
with it intentionally. You must take this burst of feelings, this unique
grouping, and engage with it in a continual, experiential contemplation.

This continual experiential contemplation begins with focusing your
awareness on the phenomenon—the plant—with which you are developing a relationship, and allowing your sensory perceptions and the feelings
that occurred in the moment of first contact to increase in intensity until
they are all that you feel.

*[There are specific] modes of perception which help in our effort to
grasp the infinite.*

— GOETHE

Allow the mood generated by the phenomenon, its emotional tone, and,
more importantly, the meanings of which these are an expression, to deepen
until your experience of the phenomenon becomes all-encompassing within
you. This deepening process requires your total immersion in the experience
of the plant itself. Nothing can be allowed to divert your attention from
your focus on the plant. Your experience of the plant becomes the only
thing of which you are aware.

During this enhanced, directed focus, your awareness will begin to
interweave with the living reality of the phenomenon, your perception to
develop into an active beholding of its ever-living, moment-to-moment
reality. Through your immersion and directed focus the phenomenon
will—eventually—come alive within you in an entirely new way. The
two of you will interweave together at an extremely deep level of

contact—your life interwoven with its life, your being shaped by its living presence.

My thinking is not separate from objects; the elements of the object, the perceptions of the object, flow into my thinking and are fully permeated by it; my perception itself is a thinking, and my thinking a perception.

— GOETHE

Your uninterrupted focus on the phenomenon and the experiential enhancement of the feeling/experience that occurred at the moment of first contact activate a deeper capacity for understanding the meaning inherent in this mode of cognition. The meanings, the communications the plant emits, will eventually emerge within you in extremely elegant and sophisticated *gestalts of understanding*. These deeper perceptions occur through a particular kind of thinking—a thinking specific to the heart—through imagination. (And not the kind of imagination you have been taught about either.)

It is the marriage of the soul with Nature that makes the intellect fruitful, that gives birth to imagination.

— HENRY DAVID THOREAU

It is not necessary to actually be in the presence of the plant when you engage in this deepening of perception.

Once you have experienced the one true thing—the burst of experiential feelings from the plant—keep it alive within you when you leave the presence of the plant. Take it home with you. Wrap it up carefully in your heart cloth. And when you have time, a time when you can be uninterrupted, when you can focus on the experience undisturbed, take it out again. Unwrap it carefully and reexperience it.

My body is all sentient. As I go here or there, I am tickled by this or that I come in contact with, as if I touched the wires of a battery. I can generally recall—have fresh in my mind—several scratches last received. These I continually recall to mind, reimpress, and harp upon. The age of miracles is each moment thus returned.

— HENRY DAVID THOREAU

Feel again that moment of first contact as if it were for the first time. Allow it to grow in intensity. Hold that initial moment of meeting within you, let all the feelings you first experienced grow until they are all that you feel. Send out in this moment a plea to the plant, to the Creator if you wish, asking to know the deeper healing properties of the plant.

We must look for help not so much to the stamen counters as to the plants themselves.

— LUTHER BURBANK

Hold this intention to the forefront of your mind and keep intensifying the moment of first contact to the limit of which you are capable. Then. . . back off slightly from it. Allow whatever might occur at that moment of disengagement to occur. Note whatever that is.

write it down if you want to

Then, before you are drawn too far down analytical paths, feel again the moment of first contact, reengage once more with the emotional reality of the plant. Enhance it, send out your plea, intensify the moment of first contact to the limit of which you are capable, and slightly disengage once again. Again, note whatever occurs in that moment of disengagement. Then repeat the process once more.

this is oscillation

The tremendously powerful enhancement of the moment of first contact combined with the earnest desire of your deepest self to know flows to the portions of the brain whose function it is to perceive—to unlock—meaning. The disengagement with the phenomenon, the slight stepping away from immediate experience, allows the brain to momentarily reactivate. The pattern of meaning within the phenomenon is then interpreted by those portions of the brain concerned with meaning; understanding then arises within you in a new form. The meanings from the plant are encoded in discrete linguistic descriptions and bursts of understanding.

Help in this process comes from the plant as well, from the environment, from some other source. Neither we nor our thinking are separate from the world.

Each emotional tone, intimation, or mood felt in response to a phenomena is an expression of meaning. And it is this meaning, or series of meanings, that you are working to turn into usable knowledge. These meanings encode deeper understandings of the plant. To pin the knowledge down, to "put salt on its tail," means deeply understanding these meanings and their patterns and then capturing them in highly sophisticated, verbal, analytical expression, in language. And that is what this process allows you to do.

This is not a forced process. Rather, the analytical capacities of the brain are allowed to generate—of themselves—linguistic descriptions that capture the essence of the thing, the meanings that are encoded within the feelings you have felt. During this process, the verbal–analytical mode of consciousness does not invent the linguistic phrases to describe the meaning of the phenomenon; instead, the linguistic descriptions of the phenomenon emerge of their own accord out of the store of memories, information, and experiences that you have accumulated during your lifetime. Here the heart and brain work together, the systole and diastole of understanding.

All my life, whether in poetry or research, I had alternated between a synthetic approach and an analytic one—to me these were the systole and the diastole of the human mind, like a second breathing, never separated, always pulsing.

— GOETHE

The heart is the primary organ of perception; the brain supplies a supportive, secondary—though essential—role.

The intellect is powerless to express thought without the aid of the heart.

— HENRY DAVID THOREAU

The brain, under the impetus of your desire and the tremendous, focused flow of information coming from the world through the heart, turns the embedded plant communications into human language and gestalt-pictures of understanding by using its stores of memories, information, and experiences.

Feel and hold this emotional tone within you, enhance it to the ultimate degree of which you are capable, then slightly disengage from your experience of the phenomenon, and

pause

You move into a brief moment in time in which you are not feeling, not thinking, not doing anything at all. This moment of nothing is highly charged with tension. It cannot be maintained for long, only a few seconds at most. If you try to hold it too long, the mind will begin thinking about all sorts of things again and you will wander off along tangents that have little to do with this process. Remember: the heart is the primary organ of perception here; the brain supplies only a secondary, supportive role. The verbal/intellectual/analytical mode of cognition used by the brain is the servant of the process. Thinking by itself will never get you to these deeper understandings.

I often say that it is best to think as little as possible.

— MASANOBU FUKUOKA

At the moment of pause, when you are slightly disengaged from the phenomenon itself, the meaning-charged experience you have been feeling with such intensity is routed to the brain for analysis. The pause, the disengagement from your feelings and experience of the phenomenon, allows the verbal/intellectual/analytical faculties to reengage. This allows the mind to generate a gestalt of understanding that captures the meanings that have given rise to the particular emotional tones you experienced. This understanding comes in a burst, a flash of linguistically encoded meaning. The knowledge emerges, seemingly of itself.

The worst is, that all the thinking in the world does not bring us to thought; we must be right by Nature, so that good thoughts may come before us like free children of God, and cry, "Here we are."

— GOETHE

Usually, the deeper understandings that are possible do not emerge within you the first time you do this. The first time (or the second, third, or fourth), perhaps nothing emerges at all. You pause, and that space of nothing is not filled with anything or, at most, contains only a tiny scattering of linguistic picture-gestalt bursts of transcribed meaning. These initial bursts are not the end point of the process, only signposts along the way. These must be added to your experience of the plant, embedded within your own heart-field, used as building blocks in the process.

However, each understanding must be emotionally compared to the living phenomenon, to the plant itself, to check it for accuracy. As you hold them within you, embed their essence in your heart-field, they will flow outward from you to the phenomenon itself (which is held in your imagination, in your memory of the moment of first contact) and then back to you. You are, in essence, holding your experience of the phenomenon within you in an enhanced form and holding as well the bursts of understanding your brain has generated, enhancing them too. You combine the two within you in your imaginal seeing. Then you slightly disengage again, and the brain makes a comparison between the two fields—between your living experience of the plant and this initial description you are holding within you. This is used to refine the accuracy of the brain's gestalt. Any differences in the two fields are separated out and these differences, these tiny discrepancies, are focused upon until the error is perceived and corrected.

The information derived through this comparison process is used to refine the results. It is essential to the eventual understanding of the plant. This refining process is work. It demands a focus of will, intention, heart, and mind. It is quite tiring in the beginning. You are refining not only your understanding of the plant, but your ability to perceive with the heart, to determine congruencies and subtle differences in meaning. You are making new muscles.

I was, indeed, then in the dark, and struggled on, unconscious of what I was seeking so earnestly; but I had a feeling of the right, a divining rod, that showed me where gold was to be found.

— GOETHE

This comparison process, while essential in refining these initial bursts of understanding, is not the destination you are seeking. After comparing (and correcting if necessary), let the linguistic scatterings you now have go. Don't make them the focus. Don't hold on to them. They are already encoded within you, embedded within your heart field, stored as memory in your brain. These insights will not be lost. You are moving instead toward a complete understanding of the plant, a burst of seeing that is far beyond these initial intimations.

You must repeat the whole process again: Attend to the plant itself once more, reexperience the moment of first contact, let it build to great intensity, send out your plea to know, then slightly disengage once more. This process must always be repeated, usually many times.

Not through an extraordinary spiritual gift, not through momentary inspiration, unexpected and unique, but through constant work did I eventually achieve such satisfactory results.

— GOETHE

You must reengage with the phenomenon and repeat the process over and over and over again until, in a burst of understanding, an articulation of the meanings occurs naturally within you in a gestalt filled with highly charged understanding, complete and whole. It is in this moment that the plant stands revealed to you, the moment when it has completely unconcealed itself. You literally see the living reality of the plant within you on the field of your imaginal vision. This living reality of the plant is not composed merely of its physical shape, its energy, engendered feelings, or those preliminary linguistic bursts, pictures, or random thoughts that you experienced. There is a depth to the phenomenon now that is far beyond those things.

You will find, however, that as you engage in this process, and before you reach the point where you perceive the phenomenon its own light, you will, from time to time, go off on intellectual tangents. This is inevitable.

the tendency to do so lessens over the years

The brain is a normal biological oscillator and is meant to be used. It is simply not supposed to be the primary location of consciousness. Because it is essential in this process and because we have been habituated to its use, our consciousness located solely in that location, it is very easy to move back into the verbal/intellectual/analytical mode of cognition and to reentrain with it.

Although the practice of thinking the phenomenon concretely by exact sensorial imagination is irksome to the intellectual mind, which is always impatient to rush ahead, its value for developing perception of the phenomenon cannot be overestimated.

— HENRI BORTOFT

Don't worry about it, just reorient yourself and continue on. When you notice it happening again, simply reengage the phenomenon, enhance the moment of first contact, and immerse yourself in the experience of the plant once more. The culmination of this process is the emergence into a unique moment of perception in which the phenomenon, in a gesture of acquiescence, unconceals itself in a burst of understanding.

There will come a time, when you slightly step back from the plant, at the moment when you are not thinking and not feeling, that the living perception of the plant as a complete whole emerges within you.

The best I could do would be to say that if one casts off everything, absolutely everything, from human thought, what emerges thereafter in one's soul—that indefinable something that one apprehends . . . that [is] nature.

— MASANOBU FUKUOKA

The organism stands forth in its own light and is understood. Knowledge of the plant as medicine (or its function in the ecosystem, or more) is directly gathered from the plant itself.

However, before the whole of the plant emerges and is revealed to the imaginal sight, there will be a moment of stasis, a moment in which the will of the perceiver and the resistance of the phenomenon to reveal itself are equal. There is an impasse, during which movement forward is difficult. You may feel as if you are pushing through cotton wool, with no forward momentum possible. At this point, it is the will of the student to know—and the depth of love he or she has for the plant—that is essential, that carries the process through. Love without the will, however, is insufficient, because there is no motive force. And without love, the phenomenon will not acquiesce to reveal itself to your gaze.

It is here, at this point of stasis, that you must hold your intention to know and not allow yourself to be distracted in your task. If you continue, to hold your directed focus, there will be a moment in which the forces involved—the will and love of the perceiver and the life force of the phenomenon—converge. There will be a moment of breakthrough where you emerge into a center of understanding. Here, the philosopher Hegel commented, "The spiritual eye stands immediately at the center of nature."

At this moment a unique experiential event occurs, filled with dynamic tension, imbued with tremendous empiric content. The student and plant interweave, each unique identity still livingly present, each tremendously potent and vivid, but interwoven together. Perceiver and perceived become linked, unified as an organic whole. Each merged into the other, their two life fields entrained.

> *Shamanic balance is not a particular stance. It is not a balance achieved by synthesis; it is not a static condition achieved by resolving opposition. It is not a compromise. Rather it is a state of acute tension, the kind of tension which exists. . . when two unqualified forces encounter each other, meeting headlong, and are not reconciled but held teetering on the verge of chaos, not in reason but in experience. It is a position with which the westerner, schooled in the Aristotelian tradition, is extremely uncomfortable.*
>
> —Barbara Meyerhoff

Once you have experienced this—once the living reality of the plant stands forth within you—you must compare it to the moment of first contact. Hold the living reality in your awareness, and hold the moment of first contact as well, slide the two together within you and observe what happens. If the living reality that has emerged within you is complete, you will experience a congruency at all points of contact of the two gestalts/images. In a sense, the waveforms, when compared, overlap, merge into one another until there is only one waveform.

this is congruency — Poetic.

With practice, this comparison process becomes extremely rapid and takes only a few seconds. It is merely a check on yourself to ensure that you are perceiving the phenomenon in its entirety.

By your constant emotional focus on the qualities—the *energia,* of the plant—the cultivation of a tremendous inner tension through enhancing the plant's emotional tone to the limits of which you are capable, by your earnest desire to know, and by your slight disengagement, the phenomenon expresses within you the essence of itself in a moment charged with meaning. And in this moment of depth perception, the moment in which the phenomenon unconceals itself, you will find yourself caught up in your living understanding of it. You will be held in a tiny moment in time, suspended in a moment of pregnant pause. And in that moment, you will literally see multiple aspects of the plant as living expressions of itself. Its medicinal actions, its purpose in the ecosystem, its relations to other plants, how it appeared in ancient times in other habitats—even what its ancestors looked like—all will be revealed.

I persist until I have discovered a pregnant point from which several things may be derived, or rather one which yields several things, offering them up of its own accord.

— GOETHE

This moment is saturated with empiric content, filled with multidimensional meaning, possessed of tremendously dynamic tension. This is a moment of highly engaged knowing, a moment of unconcealing, a gesture of acquiescence from the phenomenon itself, that allows the entity that it is to come forth and show itself in light of its own truth, to show itself *from* itself. All previous interactions with the phenomenon, up to this point, were only preliminary. At this moment, an instantaneous, living dialectic joins all parts of the phenomenon to the student in a dynamic, interpenetrating whole.

this is participatory consciousness

At this moment, the flow from you to the plant and back again becomes a living language in which nothing remains concealed from your perception. You and the plant remain two beings, yet are merged into one. You know the plant from within itself. (And, easily forgotten, the plant knows you as well.)

There is a delicate empiricism which makes itself utterly identical with the object, thereby becoming true theory.

— GOETHE

And what you know is not analytical theory, not a mental construct arrived at through linearity,

Natural farming arrives at its conclusions by applying deductive, or a priori, reasoning based on intuition. By this I do not mean the imaginative formulation of wild hypotheses, but a mental process that attempts to reach a broad conclusion through intuitive understanding.

— MASANOBU FUKUOKA

but a living reality in which the plant itself shows its multidimensionality directly to you. The "theory" of the plant is not a mere two-dimensional shadow of the linear mind, but a living, multidimensional experience of the true theory of the plant, of which its image, its form, is but one dimension of expression.

Who hears may be incredulous,
Who witnesses, believes.

— EMILY DICKINSON

From this vantage point, the mind can perceive and develop any and all aspects of a phenomenon. There is no necessity for exhaustive engagement in the minutiae of the phenomenon (e.g., plant chemistry or cellular structure) to understand it.

There is only understanding itself, from which all aspects of the thing can be known if you only direct your awareness in that direction. Here, you begin to use the imaginal thinking capacity of the heart.

In imagination I look back far into the past and inquire as to the racial history of this fruit.

— LUTHER BURBANK

The focused intention can be directed along one particular dimension of the phenomenon—its medicinal uses for instance. However, in general, there is usually a reason that underlies your deep attraction to the plant—the thing that called you to it in the first place. During the moment of unconcealing, that reason, so deeply present in your unconscious, will call forth from the plant one particular aspect of itself. The one thing that will meet your deep need, this deep call of your self. This one thing will present itself to your gaze first.

This process is not a clear-cut, easy dynamic. The moment of unconcealment is often difficult to achieve.

if it were easy everyone would do it

If the depth understanding eludes you, you must take yourself back to the initial moment of contact and allow the mood and emotional tone of the plant to once again emerge within you in all its freshness. Your intention to know, the directed focus of your will, and your depth immersion within the feelings that the thing generates in you are what lead eventually to the moment of breakthrough. You must stay with the phenomenon itself and return to it over and over again. Allow the living reality of the plant as you experienced it in the wild to emerge in all its power within you.

> *Individual phenomena must never be torn out of context. Stay with the phenomena, think within them, accede with your intentionality to their patterns, which will gradually open your thinking to an intuition of their structure.*
>
> — GOETHE

And continue with the process for as long as it takes. Years may be needed with very complex phenomena, very powerful plants, or plants that have great teachings for you, teachings that are essential to your soul's development. Things deeply valuable take time to acquire. You are seeking a deep wisdom from the world. It may grow on trees, but the harvest itself involves work. Potent aspects of very complex phenomena or very powerful plants may emerge as you do this work, but the whole may be more recalcitrant. Several expressions of a phenomenon, particular insights you gain, may seem at odds with one another, may conflict or even seem highly strange, not understandable. You must simply store them in the bag of experience you have of that phenomenon and keep working with them over time.

> *If some phenomenon appears in my research, and I can find no source for it, I let it stand as a problem. . . I might have to let it lie for a long time; but at some moment, years later, enlightenment comes in the most wonderful way.*
>
> — GOETHE

Eventually the whole phenomenon will burst into understanding within you and you will stand, suspended in time, at the pregnant point.

The focus of this work with plants creates, over time, an experientially generated database of plant medicines. Each plant will remain fresh within you, for the moment of first contact is stored away and can be recalled at any time. The knowledge of their power as medicines emerges out of the deep, living dialectic that you and the plants have created together.

All indigenous people gathered their knowledge of plant medicines in this way, directly from the heart of the world, from the soul of the plants.

All said they could talk to plants, that plants could talk to them, that the plants told them about their uses as medicines. This manner of perception, of diagnosis and healing, is the most ancient humans have known.

We do not merely engage with meanings already present in the world; we can also initiate the communicative emergence of particular meanings that we, ourselves, need. And the plants, feeling the touch of our heart communications upon them, will hear and respond.

If you desire to harvest particular medicines, plants that you already know and need for some particular thing, before you go, feel the need you have for them, and feel it strongly. Then see the plant in your imaginal vision, and let its reality enter you. When the experience is strong within you, send to the plant the need that you have. Wait and watch. Stay with the process. Keep the plant alive on the screen of your vision and keep directing to it the need you have. You will notice, if you are attentive, that the plant will suddenly shift itself. It will come alive in a particular way as you watch. It will awaken from its embedding in the livingness of the world and take notice of you and your need. Then it will begin taking into itself your communications. Stay with this process until you sense that it is complete, that everything that must be said has been said. Then send your thanks to the plant, and go to gather the medicines that you need.

When plants receive this kind of communication, they begin altering the chemicals they produce in anticipation of your gathering them as medicines. Your communications contain specific meanings, requests that initiate particular chemical responses. For the intentionally-created chemistry of plants is one primary language of response that they possess.

The more strongly the emotional reality of your need emerges within you, the more naked that need is (and the less cloaked it is in human-centered arrogance), the more responsive the plants will be and the more powerful their medicines. For this is primarily an emotional process, not a linear, mental exercise. The deep needs and values that we experience as human beings are the ones that carry the most powerful impacts, the most powerful emotional communications.

Life is no mere academic or rhetorical exercise.

In Nature's presence we are all children, nothing more, and honors and names and purses lose their significance and importance and are forgotten and only the awe and marvel in our hearts remain.

— LUTHER BURBANK

At times you may find that you need a plant that you do not know in order to meet some need within you, some suffering that you or another person have. Before you go to seek a plant for this suffering, hold your need firmly in the forefront of your mind, feel it strongly. Then send out into the world the need that you have, a request for a plant to help.

The emotional power of this need will alter the communications embedded within your heart's electromagnetic field, and they will travel out before you, powerfully impacting everything they touch. Then when you enter the woodlands, the wildness of the world, one particular plant will respond to you most strongly.

You will notice that one particular plant gathers your attention to it. It will look especially beautiful, or you will see it standing up from the plants around it, nodding to you in the sun. And some part of you will want to go to this one plant.

And it is to that one you must go, with whom you must sit, and from whom you will receive the information you need. The gesture of acquiescence that eventually comes, once you go through the process of unconcealing, will not only reveal the organism, but will also contain the information necessary to meet your need.

The power of plants flows from the deep past down through each succeeding generation, culminating in your time in this one plant that you see before you. So, too, does this way of knowing. It flows down through past generations of human beings and culminates, now, in you. When you learn this mode of cognition, take it on as your work, you carry a lineage within you that is as old as First Man and First Woman.

Today I read the description
of a medicinal plant
in a seventeenth-century herbal.
The words,
in intimate detail,
described Potentilla
and how the author used it to heal
long, long ago.

After I closed the book
and shut out the strange, time-distorted vocabulary
I took my staff,
and walked the fields
surrounding my home.
I do not know why I paused
and looked down
to see the same Potentilla
three hundred years later.

The description from the book,
like an insubstantial shadow in my mind,
arranged itself
over the five jagged fingers of Potentilla *leaves,*
his straggly stem,
swaying yellow flowers,
and clicked into place.

Wind,
blowing down
a million years of plant medicine
brushed against me.
I flickered and was gone,
insubstantial shadow in the mind of Earth.

And for a moment
I was an old herbalist in 1720,
brushing back my cloak with my hand
as I bent to look
at a plant
that Hippocrates had used
two thousand years before me.

CHAPTER ELEVEN

THE PREGNANT POINT
AND THE
MUNDUS IMAGINALIS

I am certain of nothing but the Heart's affections and the truth of the Imagination.

— JOHN KEATS

In offering the two Latin words mundus imaginalis. . . *I intend to treat a precise order of reality corresponding to a precise mode of perception.*

— HENRI CORBIN

A man born and bred to the so-called exact sciences, and at the height of his ability to reason empirically, finds it hard to accept that an exact sensory imagination might also exist.

— GOETHE

One [relies, not on the typical restrictive notion of deduction] but on a broader deductive method; namely intuitive reasoning. . . The creative roots of natural farming lie in true intuitive understanding. The point of departure must be a true grasp of nature gained by fixing one's gaze on the natural world that extends beyond actions and events in one's immediate surroundings.

— MASANOBU FUKUOKA

Tʜᴇ ʙᴜʀsᴛ ᴏꜰ ᴜɴᴅᴇʀsᴛᴀɴᴅɪɴɢ ᴛʜᴀᴛ ᴏᴄᴄᴜʀs in the moment a phenomenon unconceals itself, at the moment Goethe terms "the pregnant point," can be understood more deeply by considering a parallel process, by concentrating on the figure below.

This figure is composed of an apparently random grouping of irregular-sized black and white blotches. These are the sensory impulses that are reaching your eye, passing through you into the brain. Yet there is meaning in this picture. If you keep looking at it, of a sudden, a recognizable picture will emerge, the head and neck of a giraffe. When you see it, though, you see it suddenly, immediately.

The effect is just as if the giraffe had been switched on, like a light.
— Hᴇɴʀɪ Bᴏʀᴛᴏꜰᴛ

This instant of perception, once it occurs, is accompanied by a momentary pause.

the pregnant point

In that moment of pause, your mind is not thinking, your heart is not feeling; you are perceiving directly a specific truth that has burst on your awareness and now holds all of your attention. In this moment—when the meaning of the thing you have been attending to bursts into awareness—you are held for a moment, suspended in time, caught up in the individuality of the thing you have perceived and understood.

What happens in this instant of transition? There is evidently no change in the purely sensory experience, i.e., in the sensory stimulus to the organism. The pattern registered on the retina of the eye is the same whether the giraffe is seen or not. There is no change in this pattern at the instant when the giraffe is seen—the actual marks on the page are exactly the same after the event of recognition as they were before. So the difference cannot be explained as a difference in sensory experience.

— HENRI BORTOFT

What happens is that the *meaning* within the sensory impulses has been grasped. The hippocampus has received the sensory impulses—the image—and taken them in. It works with the relationship of the parts of the figure and its organization, and integrates the two. Then the thing that is more than the sum of the parts, the organized pattern that is there, bursts into awareness and captures the whole focus of your understanding.

The pieces that are in the picture, however, are not the picture.

the whole is always more than the sum of its parts

The meaning is in the picture, but it is not the picture. And this meaning is not merely an element of the figure. Making an exact copy of the picture will not make the giraffe stand out any more clearly to someone who cannot see it.

What we are seeing is not in fact on the page, even though it appears to be there.

— HENRI BORTOFT

This *meaning* is an added dimension to the patchwork colors on the page. It is a dimension concerned with relationships and the tension between parts and that unidentifiable something that comes when a grouping of parts suddenly unifies into one coordinated whole, when it self-organizes and begins demonstrating emergent behaviors.

We have an innate ability to perceive the unique identities that occur at moments of self-organization. It is born into us, intended for our use.

In the one simple moment when perception of the giraffe flashes into consciousness, a particular mode of cognition has been activated, one that is as natural to us as our breathing and the beating of our hearts.

Advanced educational degrees are not necessary for the use or development of this skill. In fact, they often interfere with the emergence of this mode of cognition.

Permanently.

The figure of the giraffe is, however, only an analogy for how this process works in perceiving Nature.

> *You must realize*
> *that Nature*
> *is much more complex.*
> *The actual process is much more. . . multidimensional than this.*

The picture of the giraffe is not something that itself could exist in Nature. It is a human construct, not a living reality. The dimensionality of the giraffe is much smaller—more Euclidean—than anything you will ever find in Nature. It is not real.

> *The things which enter our consciousness are vast in number, and their relations—to the extent that the mind can grasp them—are extraordinarily complex. Minds with the inner power to grow will begin to establish an order so that knowledge comes easier; they will begin to satisfy themselves by finding coherence and connection.*
>
> — Goethe

The giraffe figure is only mildly complex. It is not a nonlinear, living being. But it serves to reveal the experience of what happens in Nature when the meaning within living things is experienced directly. It illustrates the experience that occurs when a phenomenon unconceals itself, when the pregnant point is reached, when understanding bursts into awareness.

This flash of understanding, and the pregnant point pause that occurs at the moment of understanding, can be experienced with anything and everything in Nature. In the gathering of direct knowledge of the medicinal uses of plants, plants are the specific focus.

The first step is attending to the sensory impressions that come into the body from the plant. The second is attending to the feelings that these sensory impulses generate. The third is focusing on the meanings that are encoded within these feelings, the meanings that we experience as feelings.

At each step something replaces what went before. Sensory noticing replaces mental chatter. Feeling replaces sensory noticing. The experience of meaning replaces feeling.

Initially, the meanings that are experienced are simply a deeper dimension of the thing itself; they are not at this point in time unified in experience as a whole, complete understanding. The experience of the whole, living organism does not come until this burst of awareness occurs, when the pregnant point has been reached.

But the meanings that exist in a living phenomenon

in a plant

are much more complex than the meaning that is in the patchwork figure of the giraffe. There are millions upon millions of component parts within even the simplest plant. It takes only a few seconds or minutes for the meaning of the giraffe picture to burst on the awareness. With a living phenomenon, the directed focus of awareness often must occur over a longer period of time. There are an extremely large number of factors that the hippocampus must process and organize into meaning.

But remember,
the hippocampus is only a part of what is happening,
it is not you.
This is an interweaving of living beings,
not a static, linear, mental process.
It is meant to be experienced
not explained
away.

To find the pregnant point with a plant, or with anything in Nature, the moment of first contact must be anchored deeply into your experience. The focus must be the initial intimation or mood of the phenomenon or plant, the complex of primary and secondary feelings that you experienced when you first opened up to feeling.

Through the days or weeks to come, you must continually reexperience that moment of first contact, letting the phenomenon come into your awareness as if for the first time. And once it has, you must enhance the experience, letting it become all that you are experiencing in this moment of renewal. The initial intimation or mood, the feelings of the plant, must be enhanced until they are very strong in your experience. Until there is nothing else in your awareness—in your body sensations—except them.

Then, you slightly disengage, step back from the phenomenon,

and pause

letting the brain do its work of analysis.

its supportive, intended role

Then reengage with the moment of first contact once again and repeat the process.

If you keep repeating this process with anything you are interested in, that you are drawn to, that you are working to understand, there will be a moment when you slightly back off and the *wholeness* of the thing will burst into consciousness, just as the giraffe did. There will be a moment when you directly experience the depth livingness of the phenomenon within you, understand and internally hold a gestalt of its meaning, just as you did with the giraffe.

> *Intuition is the absolute-velocity insistence of the intellect upon the laggingly reflexed brain to call its attention to significance of various special-case, brain-registered, experience relationships.*
> — Buckminster Fuller

The continual focus of your intention through the days and weeks after first meeting lets the deeper parts of you know that you are truly insistent on knowing.

that you are serious

They work to find the meaning even while you sleep. At a level lower than conscious awareness, throughout each day, they work to understand.

So you can do your daily work and live your life and take only a moment or two, three or four or more times throughout the day, to let the experience of first contact reemerge into your awareness. This keeps the process going, keeps the intention to know in play, keeps the force of your desire on the phenomenon itself.

There will come a moment

always

when the wholeness of the phenomenon bursts upon you. When that moment occurs, an intense joy accompanies it, a moment in which you are caught up in the wonder of the thing itself. And there is also this pause, this pregnant point, when you and the phenomenon itself—at this moment of interwoven, participatory consciousness—are suspended in time in a state of dynamic tension.

In this moment of unconcealing when the phenomenon, in a gesture of acquiescence, makes itself known to you as a whole—the brain generates from within itself a unique multidimensional description of the plant, a form within which to hold the meanings that you now understand. It is an amalgam created from the stores of memories, experiences, thoughts, ideas, and events of your life. This amalgam is a coalesced burst of information that takes its form from things already within you. The brain's weaving together of bits of things that are already present within you puts the meaning of the plant into usable form.

> *But this assemblage,*
> *it contains something new,*
> *a thing that is something more*
> *than the parts that are being assembled,*
> *the living identity of the phenomenon itself.*

There is an immediate putting together of all these bits of knowledge into one whole gestalt, which flashes onto the surface of the mind in one tiny moment of time.

That moment, because it is so loaded with empiric content, so much feeling, so much meaning, is deeply embedded in memory. You can call it up again anytime simply by remembering the plant and the moment of first contact.

As you let the memory of first contact with a plant build in your experience, the experience of this moment of understanding will burst once again on your awareness. You will be in the pregnant moment once more.

It is from this pregnant point, during the moment of pause, when you are perceiving the multivalued, complex meanings the plant sends out from the core of its being,

> *that are its being*

that you begin to use what Goethe called the "exact sensorial imagination." But you must understand that this is not imagination as it is normally thought of, it is something else entirely.

The word "imaginary" conjures up the inescapable connotation of something unreal, something created in the mind that has no basis in reality, something made up. This definition is true only when consciousness is located in the brain, when one is using the linear mode of cognition.

Within the linear mode of consciousness, imagination is the generation of a series of thought pictures that have their relation only to the linear mode of cognition. But when the heart is the primary organ of perception, imagination is something else. It is the kind of thinking that is done with the heart. It has a particular nature, a particular form. It is a highly elegant form of seeing.

We have all heard people describe other people, in a derogatory way, as being "full of imagination." The fact is that if you are not full of imagination, you are not very sane.

— BUCKMINSTER FULLER

When beginning this process of knowing, the first step is entering Nature and focusing your awareness on the external world— directing your perception through the senses. Visually, this means that the image of the phenomenon is the object of conscious focus, and it is this image that develops into the specific kind of imagination that is important. This is the imagination that the great Islamic scholar Henri Corbin termed the *mundus imaginalis,* literally "world imagination." What he actually meant by the term was the imagination through which the true nature of the world is perceived. It is, in fact, a specific kind of seeing, a specific kind of perceiving.

It occurs after sensory focus, after feelings, after you have connected with the living reality of the plant, and after its wholeness bursts into awareness and you are at the pregnant point. In a sense, what you have inside you now is the living phenomenon itself, held in the multidimensional, imaginal field of your heart. That whole, meaning-imbued image you are now holding within you is a living reality. At the pregnant moment, once you have got over the joy—of the thing, of the process— you will find that for just a moment you are holding the image you now see within you in a momentarily static state.

You can, in this tiny instant of suspended time, view the living, meaning-filled image from multiple points of view, rotate it to see it from any perspective you wish.

When we are able to survey an object in every detail, grasp it correctly, and reproduce it in our mind's eye, we can say that we have an intuitive perception of it in the truest and highest sense.

— GOETHE

You can literally touch it at different points of contact, along multiple dimensions of meaning, with the living field of your heart. Each of these different points of touch will reveal new meanings that can only be seen from that new point of view.

The aim is to think about the phenomenon concretely in imagination, and not to think about it, trying not to leave anything out or to add anything which cannot be observed. Goethe referred to this discipline as "recreating in the wake of ever-creating nature." Combined with active seeing, it has the effect of giving thinking more the quality of perception and sensory observation more the quality of thinking. The purpose is to develop an organ of perception which can deepen our contact with the phenomenon in a way that is impossible by simply having thoughts about it and working it over with the intellectual mind.

— HENRI BORTOFT

You can, as well, literally insert yourself within the image and experientially flow down any point of orientation. At the pregnant point, consciousness can rotate its orientation and look along any axis of reality of the plant or disease. Thus you can know the medicinal actions of a plant from within the plant itself, know the disease a person has from within the affected organ itself.

You must realize my friend, that the deeper we go into this, both written and spoken words of formal language become less and less adequate as a medium of expression.

— MANUEL CORDOVA RIOS

You are literally feeling along different axes of reality, of meaning, within the plant itself. During this process you let the living field of your heart move within the living image within you. The heart, at this point, is *feeling* its way. Its touch is highly refined now, like the fingers of a man blind from birth. It is feeling meaning. Just as with eyes closed sensitive fingers can follow a rough thread along the surface of a coat, the heart now follows threads of meaning woven through the being of the plant.

> *All the faculties of the soul have become as though a single faculty; its imagination has itself become like a sensory perception of the supersensory; its imaginative sight is itself like its sensory sight. Similarly, its senses of hearing, smell, taste, and touch—all these imaginative senses—are themselves like sensory faculties, but relegated to the supersensory. For although externally the sensory faculties are five in number, each having its organ localized in the body, internally, in fact, all of them constitute a single synaisthesis.*
>
> — SADRA SHIRAZI

As the heart follows the threads of meaning through the three-dimensional image within it, the meanings your heart encounters are directly routed to the brain for analysis. Once you experience the pregnant point, the majority of these more refined meanings will be interpreted instantly, as soon as they flow to the brain.

> *It is the cognitive function of the Imagination that permits the establishment of a rigorous analogical knowledge.*
>
> — HENRI CORBIN

You pause momentarily in your feeling of this new thread of meaning, let the brain generate an understanding of it, then take the meaning and compare it to the thing itself.

> *All living things in existence have their relation within themselves; thus we call the individual or collective impression they make on us true—so long as it springs from the totality of their existence.*
>
> — GOETHE

You *feel* the meaning that the brain has generated and compare it to the *feeling* of the meaning that you are following.

> *The touchstone for what is born of the spirit [is] what an inner sense recognizes as true.*
>
> — GOETHE

The two feelings should be congruent; their waveforms should overlap. They should be a reflection of one another. A mirror image.

in a sense

During this process of using the exact sensorial imagination, you are refining the burst of knowledge that came in the moment of understanding at the pregnant point.

As you stand balanced in the pregnant point and direct your perception down one avenue of meaning, you may find that you experience a series of bursts of deeper understandings from that axis of the plant's reality. These bursts do not always constitute a whole, complete understanding of that axis or dimension of the plant. It is as if three or four of a long series of slides or photographs have been taken at random. You will instantly notice that your grasp of the phenomenon along this axis of reality is incomplete. At that moment slow yourself down, allow yourself to drop more deeply into this particular axis of meaning and focus on it. Begin to contemplate it as a depth meditation.

Take the beginning point of your seeing and the end point—the first and last slides or photographs that came to you—and begin to go through them, in order, from first to last and back again. Over and over again.

> *If I look at the created object, inquire into its creation, and follow this process back as far as I can, I will find a series of steps. Since these are not actually seen together before me, I must visualize them in my memory so that they form a certain ideal whole. At first I will tend to think in terms of steps, but Nature leaves no gaps, and thus in the end, I will have to see this progression of uninterrupted activity as a whole. I can do so by dissolving the particular without destroying the impression itself.*
>
> — GOETHE

This one axis of the plants' multidimensionality becomes your meditation now. And daily you spend time with it. As you run the pictures backward and forward and back again, the nature of the gaps becomes more apparent and, slowly, the missing pieces start to appear. Of themselves.

Each new piece is inserted into the queue and takes its place in your contemplation as you run the series backward and forward, over and over again. Each particular piece is necessarily an object of affection and contemplation.

Sink into each one, immerse yourself within it, let yourself feel its nature with your sensorial imagination. Touch it in all its details with your heart-field. Slowly, you will begin to sense just where this one piece should naturally go and how it should transform into the next image. The focus is both on the particular and the general, the whole series of insights you have received and each one individually. Eventually, the whole axis or dimensional line of the plant expression will come alive, be revealed.

Still, this is not a series of photographs

this is only a metaphor

but a living reality, flowing, complete, and always nonlinear. No gaps actually exist in the living reality that you are engaging. So you keep going with it, back and forth in your imaginal seeing, until you perceive it as one complete flow with no gaps at all. And this backward and forward process is essential, for Nature knows no direction. Nature is not linear. If you go only in one direction, you will miss essential aspects.

In this focus upon one particular aspect or axis, one particular dimensionality of the plant, you will come to know that particular expression better than any reductionist scientist ever will. You will know it from within the phenomenon itself, in all its living reality. If you are looking at particular chemical expressions, you will be able to see the electromagnetic spectrum of the sun flowing into the leaves, see molecules of carbon dioxide and water being disconnected, then recombined into new forms.

But this is not merely a mechanical process.

the world is not a mindless factory

These processes occur for reasons. Any particular chemical a plant makes is made in response to communications. Plant chemistries are a specific form of language, and they arise, they are made, in response to the plant perceiving directed meaning and responding in kind.*

*Each molecular structure a plant makes
is surrounded by its own, unique electromagnetic field
each electromagnetic field is encoded with meaning
and each can be perceived*

---

*For more on this topic see Stephen Harrod Buhner, *Lost Language of Plants: The Ecological Importance of Plant Medicines to Life on Earth* (White River Junction, Vt.: Chelsea Green, 2001).

> *by the finely tuned field of your heart,*
> *understood through the focused power of your consciousness.*

Through this method of perception, you perceive not only the chemical compound itself, but also the meaning that the plant is responding to, the communication that is giving rise to its chemical response, and you see, perceive, experience the meaning within the chemical compound itself.

You truly perceive the meaning in plant chemistry and understand experientially that plant chemistry is not pharmaceutical chemistry. The former is imbued with meaning, the latter is meaningless. One is communication, the other is noise (and not useful noise, either).

If you have a bent for molecular chemistry, the images that form within you may sometimes be molecular diagrams. If you do not,

> *it's boring, really*

you will perceive the chemical compounds through other metaphors.

> *The diagrams of molecular chemistry are only metaphors*
> *themselves,*
> *they are not real and have nothing to do with the real world.*
> *They are linear expressions, not nonlinear realities.*

And nonmolecular metaphors are usually more useful, more real, because they are less linear and less liable to put you into an analytical frame of mind in which you believe your schooling is (hierarchically) important, a state of mind in which you believe that you know something, where you put the sign in the place of the thing itself.

> *There are many dimensions in alliance space, and the pharmacological axis is only one of many.*

> — DALE PENDELL

Each individual plant makes thousands of chemical compounds, all of which are biologically active and full of impact. But these compounds are transforms of messages. They are communications. And it is not necessary to understand plant chemistry to understand their communications. A degree in English is not necessary to understand this sentence, and training in plant chemistry is not necessary to understand the language of plants. In fact, it often will do more harm than good. Grammar freaks can rarely write, for they cannot work with meaning, only form. Plant chemists suffer, as well, the same limitation.

Our ancestors, those that gathered the knowledge of plant medicines directly from the heart of the world, used other metaphors.

You should never denigrate
the form in which your perceptions arise,
never think yourself less-than,
if you do not use scientific metaphors.
Each of us must reclaim our ability
to know the world directly,
deeply and well.
(Any feeling of less-than
is merely a symptom
of the colonization of your mind.)
Don't leave it in the hands of experts.
That is how
we got
into this mess
in the first place.

The metaphors themselves are irrelevant, it is the perceived meaning that is important. And once you perceive meaning along the chemical axis, along the chemical dimension of reality of a plant, you begin to understand its healing powers as manifested along this dimension of expression. You will know that if harvested in a certain manner, with your attentive awareness focused just so, a specific healing aspect of the plant will be present in just the way you, or the ill person, needs it to be. You will use your own words for this healing aspect. It will be as real as if you called it by a chemical name

inulin

but it will be a lot more powerful as medicine if you do not.

For its power to heal is in the meaning; its chemical form serves only a secondary function. It contains the meaning but it is not the meaning, just as the giraffe figure contains the figure but is not the figure. For it is the meaning, the spirit of the plant, that heals the disease. The plant chemical merely gives it a form in which to travel. And although this form does help our bodies, form to form, we are not (solely) our physical form, and the disease is not (usually) merely a physical form, like a virus.

though sometimes, a cigar is just a cigar

The disease itself is a meaning and cannot be healed merely by supplying a form. In many instances, supplying a form merely conceals the symptoms of the disease.

the error of technological medicine

The plant chemical itself, without the meaning inserted into it, is like a word without meaning. Like saying, "I love you," to someone when your heart is closed. Such an empty phrase possesses a weight very different than one filled with meaning.

The perception of any dimensional axis of a plant, or any phenomenon, is concerned with meanings. Through attentive focus you can perceive the flow of these meanings through the organism, through its history and time, and see their uses as medicine or in healing. The more time you spend in contemplation of any one axis, the more meaning will come to you, until you have internalized completely that one axis of expression.

though there are still
359 degrees of orientation to go. . . at least

And the possible number of axes of examination are extremely large. The worlds of experience that can open themselves to your gaze are nearly infinite. The more closely you examine any axis of a phenomenon, the more you find it is like the edge of a coastline. The more closely you focus upon it, the greater the degree of your magnification; the longer and more detailed it becomes. The same is true of the phenomenon itself, as you examine it more closely in your sensorial imagination; the greater your magnification, the more numerous the number of axes.

It is exactly this process that Goethe used to understand the metamorphosis of plants—how he learned that all plants and all parts of plants were merely leaf morphed into different forms by the living power that runs through the plant.

When I closed my eyes and bent my head representing to myself a flower right at the center of the organ of sight, new flowers sprung out of this heart, with colored petals and green leaves. . . There was no way of stopping the effusion, that went on as long as my contemplation lasted, neither slowing nor accelerating.

— GOETHE

It is how Luther Burbank looked at plants and could see their history through millions of years of evolution, see them in any form in which their ancestors had ever grown. It was the source of his understanding that heredity is nothing but stored environment.

Hence the value of that search in imagination of the ancestors of our cherry in their widely separated habitats and with their widely diversified traits and habits.

— LUTHER BURBANK

It was this experience he then used to bring into existence new food plants in one or two or three generations, plants that were unique and that manifested particular characteristics that would breed true in each new generation. For he could look down any axis of orientation. He could glance at a plant and see the form of expression it had a million years ago in a different habitat. And seeing it, he would hold that image in the forefront of his mind, talking to the plant, coaxing it, until it would release that one form from within its seed, from within its stored environment. And, after planting twenty-thousand seeds, he would trot down the rows of plants and be able to tell at a glance which seven seedlings would most powerfully express just that form he had asked for from the plant.

It was in this instinct for selection that I was gifted. It was born in me, and I educated it and gave it experience, and have always kept myself attuned to it. I have particularly sensitive nerves—that accounts partly for my unusual success in selecting as between two apparently identical plants or flowers or trees or fruits.

— LUTHER BURBANK

Through this process, Masanobu Fukuoka found the true form of rice, for he talked to the rice for years, sat in contemplation with its answers, and strove to see down the proper axis of its expression to find its true form.

My method of growing rice may appear reckless and absurd, but all along I have sought the true form of rice. I have searched for the form of natural rice and asked what healthy rice is. . . If you understand the ideal form, it is just a matter of how to grow a plant of that shape under the unique conditions of your own field.

— MASANOBU FUKUOKA

It is how he found the true shape of fruit trees and learned to grow them without pruning.

If you draw a mental picture of the natural form of the tree and make every effort to [keep this form in mind in all your interactions with the tree and to] protect the tree from the local environment, then it will thrive.

— MASANOBU FUKUOKA

And it is the method he used to understand how to grow his plants without using fertilizers, or weeding, or tilling the soil and to still equal the yields of technological farming.

All of these people used direct perception to really know plants and their many axes of dimensional expression. All of them discovered the living reality of plants, developed relationships with them, and allowed them to become their primary teachers.

Here we do not set forth arbitrary signs, letters, and whatever else you please in place of the phenomena; here we do not deliver phrases that can be repeated a hundred times without thinking anything thereby nor giving anyone else pause to think. Rather, it is a matter of phenomena that one must have present before the eyes of the body and of the spirit in order to be able to evolve clearly their origin and development for oneself and others.

— GOETHE

Scientists have, by removing all meaning, intelligence, and soul from matter, caused the abandonment of this capacity of the human.

*We have been colonized
by a particular kind of thinking.*

For if the things we experience outside ourselves are only matter, dumb, unfeeling, insentient life forms—rocks, or atoms, or air— then there is no need to truly notice them. No need to merge with them in a participatory consciousness. The livingness of the world is reduced down and in this reduction the imaginal has become merely the imaginary.

> *What are these rivers and hills, these hieroglyphics which my eyes behold? . . . Why have we ever slandered the outward? The perception of surfaces will always have the effect of miracle to a sane sense.*
> — HENRY DAVID THOREAU

Reclaiming this ability necessitates personal reimmersion in the world. It means that you must *come to your senses* and feel the touch of the world upon you. And you must reawaken your heart as an organ of perception.

> *It is the heart I am trying to awaken in an aesthetic response to the world. The anima mundi is simply not perceived if the organ of this perception remains unconscious by being conceived only as a physical pump or a personal chamber of feelings. . . Awakening the imagining, sensing heart. . . cannot be accomplished without moving as well the seat of the soul from brain to heart.*
> — JAMES HILLMAN

The continual use of the heart as an organ of perception leads to refinement of the process until it becomes much more elegant and dependable than any scientific approach.

> *Deductive experimentation has never had much appeal to scientists because they are never able to get a good handle on what appears to many a whimsical process.*
> — MASANOBU FUKUOKA

When you personally understand this process, when you have facility with it and are as comfortable with it as riding a bicycle, you can use it to build a database of living knowledge of the world within you.

this is biognosis

You can build a knowledge of plant medicines similar to that developed by our ancestors long ago.

and in you this living lineage is continued

Through this process you do not think a plant medicine will do such and so, you know.

And this seeing, this knowing, can be applied as well to human disease, and the results will be just as elegant and as deep.

> *Fundamental wisdom*
> *can readily identify any and all*
> *special case aspects within*
> *the generalized whole*
> *when listening to one's intuition*
> *By which alone*
> *the generalized sub-subconscious integration*
> *of pattern cognition feedbacks*
> *are articulated.*
>
> — BUCKMINSTER FULLER

For this is our first and primary mode of cognition. The European Enlightenment was, in so many ways, merely The Endarkenment. It was the cultural moment in which the linear mind began its dominance. The moment at which this older mode of cognition was abandoned.

> *The earth is an organically interwoven community of plants, animals, and microorganisms. . . Although this flux of matter and the cycles of the biosphere can be perceived only through intuition, our unswerving faith in the omnipotence of science has led us to analyze and study these phenomena, raining down destruction upon the world of living things and throwing nature as we see it into disarray.*
>
> — MASANOBU FUKUOKA

THE FRUITFUL DARKNESS

Tell a wise person, or else keep silent,
because the massman will mock it right away.
I praise what is truly alive,
what longs to be burned to death.

In the calm water of love-nights,
where you were begotten, where you have begotten,
a strange feeling comes over you
when you see the silent candle burning.

Now you are no longer caught
in the obsession with darkness,
and a desire for higher love-making
sweeps you upward.

Distance does not make you falter,
now, arriving in magic, flying,
and, finally, insane for the light,
you are the butterfly and you are gone.

And so long as you haven't experienced
this: to die and so to grow,
you are only a troubled guest
on the dark earth.

— GOETHE

CHAPTER TWELVE

DEPTH DIAGNOSIS AND THE HEALING OF HUMAN DISEASE

O human, see then the human being rightly: the human being has heaven and earth and the whole of creation in itself, and yet is a complete form, and in it everything is already present, though hidden.

— HILDEGARD OF BINGEN

There is a time when a thing is a heavy thing to carry and then it must be put down. But such is its nature that it cannot be set off on a rock or shouldered off onto the fork of a tree like a heavy pack. There is only one thing shaped to receive it and that is another human mind.

— THEODORE STURGEON

In infinite detail her internal organs appeared on the screen of my vision. As the liver came into my sight, it was obvious from its black color that it had ceased to function and I knew that it was no longer serving to purify the blood. As this became clear to me I turned my attention to the remedy and the appropriate plants appeared in my vision—flowers from the retama tree and roots from the retamilla shrub. As the visions faded off into more general dreams I knew it was possible for her to recover.

— MANUEL CORDOVA RIOS

This way of gathering knowledge can be used with phenomena other than plants, of course. It can also be used for understanding illness and the healing of disease. The use of direct perception for diagnosis is an extremely elegant way of truly knowing, not thinking, what is going on inside the body. For with depth diagnosis, there is nothing between you and the person, between you and the disease itself.

your gaze is not bent on machines

The process is the same as that used to find the medicinal use of a plant. Only now your gaze is directed at a person. The intention is to know that person's disease, the diseased organ itself, and what it needs.

When an ill person comes to you,

take a moment and breathe deeply

let yourself relax. Now, just as you did with the plants, let yourself notice this person. Focus your *senses* upon him.

You will see the person first with the physical eyes. And he will appear a certain way. His skin will have a specific quality. It will have a life (or lifelessness) of its own. A color. A texture. An aliveness or deadness. A degree of health or illness.

Do you like it or not?
Are you drawn to his skin or repelled by it?
Would you like to touch it? Or not?

The clothes he has chosen—these will have sensory qualities, too. Their colors and textures. Their combination.

let your gaze focus on them now

Let the visual reality of the person's appearance enter you. Let your visual senses completely focus on his appearance. Notice everything, and notice, as well, how you respond to what you see. For this person is talking to you through his body, and you can perceive everything he says.

if you want to

The person will also be talking to you in words, of course, getting comfortable with you, with this new place he is in. Trying to establish a relationship. Trying to see if you will really be present with him, really receive his pain. Wanting to know if you can help him.

As you listen, pay attention to his voice. Allow yourself to really hear the sound of it, hear the many communications within it.

How does his voice sound? What is its timbre? Its intonation? Is it musical, or monotone? Living? Or lifeless?

Do you like it? Or not?
Are you repelled by it, or drawn to it?
What is the primary feeling
that comes to you
as you listen to his voice?

As his communications impact you, you will begin to respond in numerous ways. You may have odd, stray thoughts that pop into your head. Your body may shift its functioning—your breathing become rapid and shallow or slow and deep. Your forearms may feel tight or your stomach begin to hurt. A wide variety of emotions may emerge. Attend to all of these things, for all of them are information about the person you are meeting.

These initial intimations are the beginning of knowing this person. But just as with plants, you can take this deeper. You can work to see the living reality of the disease itself. You can seek the pregnant point. Now,

focus more directly

what about the person do you notice most strongly? To what part of the body is your awareness drawn, seemingly of itself?

It could be her lungs, her hands,
or the way she holds her mouth.
It could be his pelvis, or her breasts.

Let your awareness center on that part of the person's body.

How do you feel seeing it? Are you willing to follow your awareness where it leads you? Are you willing to really feel what is there?

Let the feeling that this part of the person's body engenders become strong within you. Enhance it until it is the strongest thing you feel.

You will have particular responses at this point. It is important to notice them, important to not allow social politeness to get in the way of your seeing. We are taught to not look too closely at people,

because it's rude

or to peer too deeply into them. You must suspect any mental restraint on your part, any thought that tells you not to see, or that you are bad for seeing, or that attempts to justify how they are in order to lessen the emergence of these observations into your awareness.

You must stay with the feeling that has emerged into you from the part of the body to which you have been drawn. This is the one true thing that you must stay with throughout the process. The part of the body that you are drawn to first is the part that is making the greatest communication to you, that you have already begun communication with, the part that is calling out most strongly to you upon your first meeting.

Any aspect of the phenomenon will ultimately reveal the whole phenomenon itself, will allow your entry into its living reality. The part that draws your attention is the doorway that is most open to you. It is with this axis of the phenomenon's multidimensionality that you must start.

The emotional tone of that part, its intimation or mood, is the feeling of the specific communication it is making to you. A primary emotion will be associated with that part of the person: mad, sad, scared, or glad. There will also be a series of non-linguistically identifiable, secondary emotions that have tremendous impact but are not so easily put in words. This complex of secondary emotions will have a specific, overall feeling.

just as a bouquet of flowers has a specific, overall appearance

Keep all these emotions in mind. Hold them to you, anchor them into your experience, wrap them up carefully in your heart cloth. For you will need to return to them over and over again.

You will find that you can continue talking to the person with a part of your consciousness while all this is happening. You will find that your ability to do two or more things at the same time grows the more you practice.

To receive a simple primitive phenomenon, to recognize it in its high significance, and to go to work with it, requires a productive spirit, which is able to take a wide survey. . . And even this is not enough. . . uninterrupted practice being still required. . . it is necessary to be constantly occupied with the several single phenomena (which are often very mysterious) and with their deductions and combinations.

— GOETHE

While you are talking, allow yourself to drop down into the complex grouping of feelings you are experiencing. They have a specific identity. They are an ordered grouping that, together, makes up the communication you are receiving. Let your experience of this complex of subtle feelings become stronger, until it is all that you feel.

Enhance the mood or emotional tone. Let it become so strong that all other things fall away from your awareness.

(Say you are drawn to the person's chest first; perhaps it feels sad to you, compressed, closed in. In addition, a subtle combination of emotions is emerging from that part of the body, like a complex blend of odors along a meadow trail. Keep all of these in the forefront of your mind. Strongly enhance them.)

Now. Let yourself begin to care for this part that has drawn your attention.

If you have a resistance to caring
this is crucial information.
Do not let yourself be distracted
by feeling bad
about not wanting to care for her
Work with yourself
so that you can care for her anyway.
If you do not think you can truly come to care for her
(this is the second question you must ask yourself)
then don't work with her.
(The first question must be: "Is this person meant for me?")

Allow the natural capacity you have for feeling love to come into your awareness. Feel the field of your heart and let yourself be aware of it as an organ of touch. Fill it with the caring that you have. Expand the field of your heart, like hands now, and envelop the part of the body that has commanded your attention. Touch it, be sensitive to its every nuance. Allow its communications to come deeply into you and send your caring back to it in response. Cradle this part in the touch of your heart-field, soothe it as you would a little child. There may be things that you spontaneously sense you must say to it. Say them silently, within yourself; fill these communications with caring, and send them out through your heart-touch into the part itself. It will respond to your communications, and you must respond to it again in turn. Engage in a living, meaning-directed dialogue with the part. It will begin to soften, relax to your embrace, allow itself to be held.

You will have to truly care and genuinely be present.
You must be real
for it to be willing to respond to you.

Enhance this experience so that it is all encompassing. Let your caring flow deeply into that part. Take as long as you need to to complete this part of the process.

When you feel that it is right to move on, keep holding onto this living experiential dialogue, but now let your greater intention—the emergence of the illness and affected organ system as a *whole* phenomenon—come into your awareness. Add your desire, your request, to know fully the organ system and its disease to the energy flow you are directing toward that body part.

[There must be a] striving of the human spirit to make a whole of the object it observes.

— GOETHE

Let yourself drop deeper now, into the organ system that is *under* the part of the body you are drawn to. Let yourself drop down into the person's lungs, for example, if it is her chest to which you have been drawn. See, through the sensorial imagination of your heart, the lungs before you.

Sometimes you will encounter resistance at this point—perhaps from yourself, perhaps from the person, or even from the organ you are trying to see. It doesn't matter where the resistance comes from. Just note it and let it pass through you. Shunt it into another part of your mind for analysis, then let it go. Allow that other part of your mind to work on the nature and meaning of the resistance, but keep your primary attention focused on the person's lungs. *Intend* to see.

The organ will begin to emerge in bits and pieces. Short flashes of vision, specific feelings, linguistic phrases. *Will* the lungs, implore them, to emerge before your sight. Touch them, and touch the bits and pieces of insight that have emerged, with your caring, just as you did the part of the person's body that drew you. Let your caring flow deeply into them.

They are a living being.
A unique self-organized organism

with emergent behaviors
and intelligence.
They have a particular identity,
one that you can feel.

Sometimes, in this work, you may see a picture of the organ you are striving to know. This may be merely a picture of the organ that you once saw in a book. Because you know how the organ looks, you instinctively brought that picture forward into your awareness. This can become a focal point. Keep this image before you and work to enhance it, to bring it alive, so that it becomes more than just a remembered picture, but instead is the living reality of the organ itself. Direct your caring strongly into the image and will it to come alive, to emerge into your awareness as a fully living phenomenon. Ask the organ to reveal itself to you.

Build all these feelings, pictures, and your intention to know to an intensity of awareness, until it is all you feel. Now slightly back off. Disengage and pause. Notice what comes into your awareness. Incorporate this into the work. And repeat the process.

This may take some time, for unconcealing is often not immediate. During this initial work, you will get a sense of the organ system and what is wrong, a general intimation of the problem, and this is where you must start. The intention, however, is to know intimately, to completely know in the depths of your being what is actually happening in the organ system, to know what the disease really is, not merely its symptoms.

It does help
if you get the person to send you
a list of her symptoms
before she comes to meet you
for the first time.
Your relationship begins the minute you receive it,
the minute you begin to feel into it, to think it over
in this new way of thinking.

After the person leaves your presence, each day thereafter—three or four times a day—in your imagination, you must return to this place, feel once again the feelings, see once more the organ system in front of you, and then repeat the process, just as you did with the plants.

Attempt to pierce the veil between your understanding and the organ system so that you fully see what is there. This can take anywhere from two minutes to two months, depending on your skill, the disease itself, and the strength of your intention.

Medical diagnosis often takes as long.
Or longer

You cannot power your way through. The drive to know, the intention, is crucial, but it is not the only factor at work. You are establishing a relationship with the person, her organ system, and the disease complex. It is an act of intimacy that is extremely deep.

breathe, be patient,
allow this to be a contemplation

During this process, your being will begin to interweave with hers, to interweave with her organ system, with the disease itself, until there is no boundary between the two of you. The only part of you that will

and should

remain distinct and unique is your awareness, your perceiving consciousness. But your feelings, your livingness itself, will interweave with those of the person you are working to understand.

a deep empathy occurs
it is inherent in this work

During these periods of concentration on the illness, the person, and her organ system, everything else must leave your mind. These things must become the sole center of your attention.

You are feeling the electromagnetic spectrum of encoded energetic information that the phenomena emits as a specific grouping of feelings. These feelings must be enhanced. You must return to them anytime you become confused or lose your way. The more you enhance these feelings, this gesture of communication from the person's body, the more your brain can work to translate the communication.

So, hold the experience in your mind, step slightly back from it, then reengage it. Always asking, "What is truly wrong? What is happening to you? Let me see you."

Eventually there will be a flash of understanding. A moment in which the organ stands forth in its own light, in which it unconceals itself and appears in front of you, revealed.

you will see its living reality
on the field of your imaginative vision

The organ system and its disease will be clear to you then. You will be able to see the effects of the disease on the organ, how the organ is transformed by it. You will know whether it is an ulceration of the gastrointestinal tract, cancer, inflammatory bowel disease, parasites, or a bacterial disease that is causing her bloody stool.

You will have, as well, a number of experiential responses to the organ system you are perceiving and to the disease itself. Many organ systems, when they appear, possess a specific kind of unaliveness, not so much from their illness as from our historical (non) relationship with them. Our bodies have been tremendously denigrated. The result is that our organ systems, when they appear in detail, will often appear diminished, not alive and vital. If you repeat this process with a wild animal—not one in a zoo—you will perceive an organ system that is very different. It will be alive, vital, intelligent, perceptive, and aware.

the organ system of a healthy, wild animal
will emerge in your inner vision in its true form
as Nature intended it to be

Perceiving the unaliveness of an organ system can be a bit frightening. It is a reality we do not often encounter.

consciously

When the organ appears on the screen of your vision, work to bring it alive. Talk to it, feel deeply into it, exhort it to awaken, to emerge in a more vital form. Organ systems, under the press of our habitual notnoticing and not caring, from our denigration of the body, lose their inherent vitality. They become unconscious, like a repressed part of the personality.

This method of seeing, this mode of cognition, these directed communications, calls them back into the world, awakens them, and helps them become more vital and alive.

When someone truly sees us
and, in caring, urges us
into the warmth of a loving embrace,
we leave the darkness
in which we have taken refuge

and come once more
into the light.

At the moment the organ system unconceals itself, its actual reality will come into you. It will emerge *inside* you, and your consciousness will shape itself around it. You will simultaneously experience the organ system from both your and its point of view. When this happens, your orientation, your experience of life, the way you see the world, will be altered, for you are seeing and experiencing through another organism's perspective. This may be disorienting, because an organ system sees life from a very different orientation.

The closer the reality of the organ system is to your normal point of view, the easier it will be to experience its reality. The farther away it is, the more disturbing it will be. Experientially assuming an extremely different orientation will in itself teach you about the narrowness of your normal point of view. The resistance you have to this new and very different perspective is information about how and how much you cling to your normal perceptions.

When you take on a phenomenon's point of view, you are thrust into new terrain, and you must learn that territory, overcome your fear of it, and genuinely look at the new world in which you find yourself without preconceptions, without personal emotional bias. You must learn its landmarks and signposts, and be able to respond to the person whose world you have taken into yourself as a companion traveling the same path. The person (and the organ system) will know in that moment of unconcealing whether she is companioned or judged, accepted with open heart or in the presence of someone who fears her internal world, just as she herself may.

> *The spirits we call on "know our minds" and if they find our conviction faltering they will not heed us, nor the words we speak.*
> — SWIMMER

Your own lack of fear is itself one of the primary components in the healing. To truly work with the living reality of an organ system, to flow down its different axes of meaning, to merge into any of its orientations, you must work within yourself to overcome discomfort with and fear of the new, and very different, point of view it presents to you.

you must enter the fruitful darkness

One of the best ways to do this is to sit in contemplation with the organ system every day, until you are comfortable with it and your new way of seeing and experiencing it. At the same time, practice moving yourself along different axes of rotation of the phenomenon, just as you did with plants. You will come, in this process, to be comfortable with each axis of the phenomenon, to know its interior terrain as well as you know your own. You will become, in the end, its intimate friend.

As you move down each different meaning-axis of the organ, you learn multiple aspects of its nature. You can see the history of the organ system before you, experience the emotional realities that gave rise to alteration in organ function. Literally see the emotional bodies of both the person and the organ system before you.

If you wish to see how the disease began or how long ago it started, you can. If you wish to see the effects of the disease on other organs, you can.

> As soon as we observe a thing with reference to itself and in relation to other things, foreswearing personal desire or aversion, we shall be able to regard it with calm attention and form a quite clear concept of its parts and relationships. The further we continue these observations, the more we are able to provide links between isolated things, and the more we are able to exert our powers of observation.
>
> — GOETHE

The disease itself, as you will discover, has a particular identity. It is perceivable as an intelligent entity in its own right. It will have its own desires, force of energy, and reality.

> *it, too, is a self-organized system with emergent behaviors*
> *like a grasshopper*
> *or a virus*

You may also feel, as you work deeply with the organ system and the disease, the pain the person is experiencing. It is important that you receive this pain, establish a relationship with it, and not turn away from it (or the person) because of the impact it makes or because you fear it.

Notice, too, how you respond to the disease and pain that you encounter. Your response to the illness, the pain, is essential information. Some pain is welcoming, some frightening. The *way* you naturally respond to it will tell you much about its nature.

just as it will tell you about your own

As you come to understand the nature of the disease and the pain that is present, you must begin to establish a dialogue with it, just as you did with the plants, just as you did with the organ system itself. This communication is essential. Understanding the disease and the pain, seeing them in their wholeness, is just as important as seeing the organ system in its wholeness.

Understanding them in their own light without preconception or bias, without labeling them an enemy, is essential. For diseases are as much a part of the Earth as we are.

Nature sees both cereal grains and weeds, and all the animals and microorganisms that inhabit the natural world, as the fruit of the earth.
— Masanobu Fukuoka

They have often been here longer, will remain long after we are gone, and they have larger ecosystem functions than merely making our lives difficult. They need to be seen in their wholeness.

To give an example, when an insect alights on a rice plant, science immediately zeros in on the relationship between the rice plant and the insect. If the insect feeds on the juices from the leaves of the plant and the plant dies, then the insect is viewed as a pest. The pest is researched: it is identified taxonomically, and its morphology and ecology studied extensively. This knowledge is eventually used to determine how to kill it.

The first thing that the natural farmer does when he sees this crop and the insect is to see, yet not see, the rice; to see, yet not see, the insect. He is not misled by circumstantial matters. . . What then does he do? He reaches beyond time and space by taking the stance that there are no crops or pests in nature to begin with. . . This insect is thus a pest and yet not a pest.
— Masanobu Fukuoka

Through this process you enter the territory of disease and healing, weave with it as a participatory consciousness, holding yourself at the fulcrum point, and become the channel through which resolution can occur. There is, in consequence, a deep necessity for you to learn to not fear this territory, to learn to walk within it without letting your fear stop you.

it is always frightening
to enter this fruitful darkness
but that is no reason to stop,
it is just part of the process

There will sometimes be a tendency, too, when empathy is being established, for your body to entrain with the person's, to take on his disease complex, his physical patterning. You can, if you know your body well enough, allow this to continue until your body itself is (temporarily) ill in just the same manner. Then, by an internal examination of your body, you can see exactly how the disease is manifesting and what it needs in order to be healed. And knowing your body so well, you can then allow it to return to its natural functioning.

this is a hard, and sometimes painful, way to do it
but some of us naturally incline to it

If you do not know your body extremely well, this approach is not necessary; it can be distracting. So, you must notice this immediately,

without fear

step back slightly, and reorient yourself to your normal mode of functioning.

the point now is for the person to entrain with you

As you establish rapport, she can also entrain with you, her body model its functioning on yours. Her waveforms will adapt themselves to yours. Her body will begin to follow the lead of your physiological systems.

just as your biological oscillators
entrain with your coherent heart
her biological oscillators
will entrain with your coherent heart

If you spend many years training yourself in this process, you can also come to learn the true form of each particular organ system—just as Masanobu Fukuoka learned the true form of rice. For each organ system has its own identity, an identity that exists independent of its current manifestation in any particular person. While an organ adapts itself to the particular field in which it develops, it has an identity that lies underneath and within the world. From this any individual organ system takes its original patterning.

just as plants take theirs

Once you know the true form of an organ system, whenever you encounter it you can see exactly how it has malformed. (It is this malformation that has allowed the "disease" to emerge as it has.) In comparing the organ's true state with the one you now see in front of you, you can determine just where things went wrong. Then, if you bring the power of Nature directly into the organ system, it can begin to restructure itself once more into its true form.

you can also ask the organ system itself
what it needs

When you have developed facility with this kind of diagnosis, each of these potential approaches will be incorporated within you. The diagnosis will then be much quicker and easier. Sometimes, the organ system itself will simply appear on the screen of your vision and you will immediately know what is wrong. For you will have learned the territory well, and you will automatically process the information, separate out, and clear your internal responses at a level below conscious thought.

Once you achieve the pregnant point with any disease or organ system and it has given you its gesture of acquiescence, revealing itself to you, send out from your self the request for a plant or series of plants to help. Hold the living reality of the organ system within you, allow its need and the need of the person to emerge, and hold your understanding of the disease within you as well. Let all of this grow in intensity, and when it reaches a peak, send up a request from within you into the world, for help, for healing. In that moment the database of living plant knowledge within you will offer up the one or ones that will help.

a plant will come to you then

You will suddenly become aware of a plant standing near to you. (Sometimes this will happen immediately, in the moment you see the organ system before you.) You will feel its vibrant life force coming from the living image that you see. If you now ask for proper dosages, how to prepare the plant for medicine and how much to give will then enter your awareness.

When you begin to prepare the medicine, hold the plant's living vibrancy and intelligence in the forefront of your mind. Speak to it at every step of the way; keep it alive in your experience. It is important to not switch into the verbal/intellectual/analytical mode of cognition at this point, for you will kill this living process with the word.

Once the medicine is prepared and you have given it to the person, you must hold that person, that living being in front of you, in the forefront of your mind.

At the same time, and in the same way, hold the living medicine. See too the living reality of the unhealthy organ in front of you, its illness. Be aware, if you can, of the true form of the organ, and direct this true form to flow into and through the diseased organ.

Bring the medicine, the organ, and the disease together; see them touch in your imagination and begin to flow into one another.

> It is the power of Nature to heal, not that of men.
> In reconnecting the living reality of the organ system
> to Nature and the plants,
> it is Nature and the plants
> that do the work.
> They teach the organ system
> how to be and what to do,
> teach it how
> to return
> to its true form
> and be healed.

You must see all of this in your imagination each day that the person is taking the medicine. Each day, as the healing progresses, you will be able to actually see the disease healing. You will know the moment in time when the healing is done.

[Man] is neither in control nor a mere onlooker. He must hold a vision that is in unity with nature.

— MASANOBU FUKUOKA

This aware focus on organ systems allows each one of them to emerge out of the living system in which they are embedded. You will develop a living dialectic, much like the one you developed with plants, and the organ system will speak to you in its own mode of representation, standing forth unconcealed. From this mode of cognition you will know what is wrong with any particular organ system; you will not merely *think* it.

This knowing will itself be communicated to the person you are helping and plays no small part in his or her healing. Further, the elegance of understanding that comes from this mode of cognition far surpasses anything that can be generated by the verbal/intellectual/analytical mode. With the verbal/intellectual/analytical mode, elements stand out, relationships become only a shadowy background, barely perceived. With the holistic/intuitive/depth mode, relationships, intercommunications, and interdependencies are vivid. The psychological and spiritual elements of disease stand out sharply, along with the physical. There is no separation between them. They are only different facets of the same thing, different axes of its dimensionality.

You must understand, however, that this approach to healing is not a technique. It is not a reductionist series of steps, of specific behaviors, like planting and watering a lawn. It is a communication.

The steps I speak of are only a map of the territory; they are not the territory itself. The important, crucial thing is to be in the territory itself and learn how to confidently find your way, one congruent, feeling-step at a time. Each interaction, each living communication has its own identity, its own territory. Your sensitivity must be trained to the point where not the slightest thing escapes its touch, where even the slightest alteration of meaning or emergence of new meaning captures your attention. You will then generate new communications in response. You are in essence perceiving meanings, working to understand those meanings, and crafting responses that are as deep and meaningful as the communications you received.

These skills you are learning are only the framework of the process, without the living flow of soul force that you bring to it, without the flow of meaning, they will accomplish nothing.

The teachers told us quietly that the way of experts had become a tricky way. They told us it would always be fatal to our arts to misuse the skills we had learned. The skills themselves were merely light shells, needing to be filled out with substance coming from our souls. They warned us never to turn these skills to the service of things separate from the way. This would be the most difficult thing.

— AYI KEWI ARMAH

Thus, the essence of this work must never be mistaken for a static form that can be repeated or mass-produced. It must not be used merely for resource extraction from the nonmaterial realm. It is not a one-size-fits-all process.

this is not medical school

This is a participatory process in which you must be livingly present. It is based in a deep intimacy. There is no place here for a disinterested observer

someone standoffish

There are no degrees here, no hierarchy,

In the human spirit, as in the universe, nothing is higher or lower; everything has equal rights to a common center which manifests its hidden existence precisely through this harmonic relationship between every part and itself.

— GOETHE

no better or worse, no *professional* experts. There are, instead, two human beings, livingly present with each other, establishing an empathy of connected understanding. You may know more (about this process), but you are not *worth* more than the other person. You both have the same inherent value, suffer the same predicament, possess the same intrinsic nature.

the sufferer is your teacher
and you must be livingly present
to learn why

There is a particular feeling to this kind of living communication, one that you will come to recognize with practice. The entity that is the living communication flowing between the two of you must be carefully nurtured and enhanced. For it is this substance that is the most important thing. The stronger this substance, this feeling, becomes, the more strongly you will be linked with the person who has come to you, and the more powerful your communications will be.

The living dialogue in which you are engaged has its own balance point, just as riding a unicycle does. With practice, you will be able to tell immediately if the balance is shifting, immediately understand why (if you are attentive enough), and craft a response to restore it. This can happen only if the communication remains a living, shifting, nonlinear, self-organized entity.

Communication is a process of perceiving, interpreting, crafting, and directing meaning. And in this process, you are training yourself to perceive extremely subtle meanings and to craft others, just as subtle, and then to direct them and be able to perceive when they are received. As soon as a communication is received, the person or organ system will alter its orientation.

When something has acquired a form it metamorphoses immediately into a new one. If we wish to arrive at some living perception of Nature we ourselves must remain as quick and flexible as Nature and follow the example She gives.

— GOETHE

The specific identity with which you have established a relationship will slightly change itself through some medium of perceivable expression. So, you must keep coming back to the phenomenon itself, see it anew again and again. For it is a living entity and it changes itself continually.

Nature is a fluid entity that changes from moment to moment. Man is unable to grasp the essence of something because the true form of nature leaves nowhere to be grasped. People become perplexed when bound by theories that try to freeze a fluid nature.

— MASANOBU FUKUOKA

For this kind of healing to be effective, you must be livingly present in the process, intimate with the person, organ system, and plant. You must really believe the things you communicate. You must really feel the caring you communicate. The most important thing is to respond with as complete a congruency as possible—that is, with all parts of you agreeable to the process and with what you are doing. Your body must congruently reflect all your communications, your unconscious parts must congruently reflect all your communications. No part of you can be held back, surly, fearful, or uninvolved.

To be able to respond with no self-delusion or hiding or incongruency, you must really like the person, including the part that is ill.

just as you must care for the disease itself

This interaction must be genuine, it must be real. The deep self, anything from and within Nature, and most especially plants and organ systems, know bullshit when they encounter it.

this is a challenging place to walk

To achieve elegance, you must not only develop skill in the process, but also a rigorous self-examination. Your perceptions must become as free of personal, psychological unclarities as a still pond is of ripples. There is no magic in this, but there is much skill. To develop this skill you must become highly sensitive to the living field of your heart, be able to notice any alterations that occur within it as it comes into contact with other living fields, and figure out just what each of those alterations mean.

There is nothing magical or mysterious about my methods, and what I have learned to do others can learn to do, and what I have started others can finish, and what I have learned about the laws of Nature can be applied by others and added to by others, if only they will waken to the possibilities that exist.

— LUTHER BURBANK

INTERLUDE

The woman who had come to see me was tentative at the door, hesitant. Her eyes were nervous, quick, surrounded by lines of worry. She eddied in the door like a wisp of smoke, whispered across the room, and hovered lightly in the chair. She was forty-five years old, short, thin and wiry. Her skin was pale, washed out, her hair a brown, not-flowing shadow of life. Just there.

She had come because she could not breathe. She had asthma.

I said hello as she sat, began drinking tea, telling me her life in many languages. In words. In the small flutterings of her hands. In intonations, the rise and fall of her voice as she spoke. In the slight shifts of her body, in the tiny patterns of emotion that crossed her face. The *shape* of her body. The clothes she wore.

Her asthma had come on suddenly with no prior history. It had been almost twenty years now. Her medications were many, expensive. Laden with side effects.

I responded to her gesturings of communication. Talked with part of my mind

hearing her speak of her life

while another part looked deeper, seeking the path the disease had taken in her. Searching for traces of its truth.

Her chest caught my attention, standing forth of its own accord. Beckoning.

My attention centered there and I breathed into it, letting my awareness move deeper, touching its shape. *Feeling* my way. I felt a sadness come over me, an overwhelming urge to cry. And then my chest began to feel tight. The muscles clenched, closed down. I began to hunch over slightly, curl around myself. My chest hollowed and I began to breathe high up, rapidly, in small quick bursts of breathing. My breathing a tiny bird, fluttering against the walls of my chest.

I began to feel afraid then, slightly hysterical.

I calmed myself, breathed more deeply. Sat back in my chair. Felt a wave of relaxation flow through my muscles. Slowly, one by one, they unclenched.

I let myself care for her then. Sent out a wave of caring from me to her. Let it touch her chest, hold it in the hollow of caring hands. Waited. . . waited. . . waited. Breathing slowly, softly, calmly. Into her chest. Slightly urging it, slowly. . . slowly. . . slowly, to relax, to calm down, to breathe.

It took a few minutes.

I saw her sink more deeply into the chair, her muscles beginning to relax. Her skin tone was changing, the muscles and the skin itself softening. Her face relaxed. And she took a deep breath. There was a slight wheezing sound. Then she took another, and deeper, breath. Her chest began to open up slightly, the muscles letting go.

And all the while, of course, we were talking. I let my deeper breathing flow into the tone of my voice. As my breathing deepened, slowly my intonation deepened with it. My words, originally dancing and quick, in tune with her breathing, began—patiently—to slow down, deepen, and become more calm.

Her eyes softened. Grew moist. Slightly unfocused. She began to tear up. Softly. Silently.

A few tears trailed down her cheeks.

My gaze focused on her chest, and I began to embed communications in my talking, telling her chest it was okay. That it could relax, breathe, tell me its secrets.

Her talking began to keep pace with my own, grew slower, more studied, less nervous.

She took a slow, deep breath. Smiled hesitantly. Her skin began to get some color, to glow a bit. I smiled too, then, and nodded slightly. I let my voice wrap her up, hold her in its arms. Telling her it was good, that she was okay now. Telling her lungs that they could relax.

I let my awareness flow deeper then. Through the surface of her chest, into her lungs.

My lungs seized up. It was hard to breathe. I couldn't get my breath. Some small part of me was afraid. Hysterical. I turned a part of my vision, focused it inside me. Looked around. Saw the frightened part of me and held it in my arms. Soothed it with soft words, my intonations and presence saying more than words ever could. The frightened part of me began to relax, feel better. Not abandoned.

Keeping part of myself touching my own lungs and that frightened place, I turned my attention back again to her lungs. Letting myself sink deep within them, I began to *look*.

There was a slight hesitation, like pushing against a mushy blanket, soft cotton wool. A resistance. I sent my caring into the resistance and deeper, through it, into her lungs. Asked them to let me see. I stayed present, *breathed* into the experience. I enhanced the strength of my caring. Looked deeply, focused my seeing, *wanted* to see.

There was a slight hesitation again. Then a sudden movement into a still, quiet center. And I could *see,* could *feel,* the living reality of her lungs.

Their color was off. Some strange cross between gray and mucusy-white. It was old mucus, an unhealthy brownish-yellow glue. My nose wrinkled slightly as the smell came to me. It was barely there, slight, as faint as the whispers of children. A sick smell, my stomach nauseous to its touch.

The surface of her lungs, the cells themselves, were clogged, gunged-up. *Suffocating.* Grayish, not pink. Covered with a blanket of old gummy sludge, a blanket that wove over and into them, through them. The mucus was dead, *unliving,* not like normal mucus, which is thin and watery. A glowing, moving, living thing, shining out its healthy life.

Hers was dead, unmoving, held in place. Old and unattended. Its life force was gone. The cells of her lungs, the tissue itself, were taking on that deadness, that unhealthy unaliveness. The lungs were slowed down in their function, held back by this *oldness.*

My attention *focused* on the lungs, my seeing alive to every part that had been revealed to me, I reached into her lungs with my caring then. Directed the living, feeling field of my heart to hold them, envelop them. My caring moved deeply within her lungs, interweaving with their tissues, *holding* them, all of us now suspended in a living moment of time. Then, still holding them, still present with them, I turned a part of my attention at a slight angle, sent it out into the world. Sent out a request for help, a prayer from my deepest being, my earnest need flowing out through this channel I had opened into the world. At the same time I kept a living channel open through me into the living reality of her lungs.

Then I felt the *need* of her lungs and let it flow through me and out, attaching it to my prayer for help so that they flowed together, interweaving with one another, flowing as one earnest need and plea.

I felt that living communication flowing out, its field spreading wide, touching the living reality of the world. I felt the living intelligence there, deeply embedded in its own work, its own living. Then, as it felt my touch upon it, and knew that it was genuine, that there was caring behind and within it, it quickened, awakened, *turned* toward me, and *saw*. A living flow of energy then came back through the channel I had opened between us, a flow with caring in it. A deep caring and loving came back from the wildness of the world. The world from which all of us have come.

> And into my mind flashed an image of skunk cabbage
> powerful, green,
> luminescent in wetland forest.

I relaxed my touch then; my concentration softened. My *focused* awareness let go. And still talking to her with that other part of my mind, I came back into the room, and let these new understandings flow into my talking. I began to weave the healing of this plant medicine into her body, into her life.

Later, when talking was done, I gave some of the skunk cabbage tincture to the woman with asthma, put only one drop of it on her tongue, and watched as she closed her mouth

> closed her eyes

and tasted it. Watched as she let her system absorb it. I saw her breathe deeply, then open her eyes suddenly as it penetrated farther into her. Then I saw the smile that came and watched her body relax, the tension lines smooth out of her face.

I remembered her lungs then and found myself standing once again with them, their living reality before my vision. The color of the mucus was present once again, its *thickness* within my awareness.

Holding this in my vision, I turned slightly, opened a channel from me out into the world, into that wetland forest, to skunk cabbage. Saw the plant once again before me, her roots gleaming wetly in my vision. Saw and felt the living reality of them, felt once more her medicine.

> let her name arise within me

Then, calling to her, asking her to come with my caring, my prayer went out to her, touched her, and she turned, awakening. Her living reality came into my awareness and flowed down that channel of communication, into this woman's lungs.

Then I gave the woman a full dropper of the tincture.

I could see the tendrils of the plant, her roots

their color matching that of the lungs

begin to wriggle into the lung tissue, inserting themselves deep into the woman's cellular tissue, *flowing* through them. I saw the plant interweaving with her lungs. The power of the plant, the medicine of skunk cabbage, flowed deep into her cells, filled her lungs and the mucus began to thin, the color of the lungs, to change. The mucus became more watery, began to run, to flow. It began to ooze out of the tissues, the cells clearing. And I could see the healing begin, see the plant teach her lungs how to be. I could see the lungs begin to take on the power and strength of the plant. Then, from the Earth came a power older than the human, flowing up through her body and into her lungs. An ancient power, old, deep, dark, and silent. Her body began taking it in like a food that she had long forgotten. Her lungs reached out to it, relaxed into it,

settled down

into it. The power began to flow up into her lungs and out into the world. The old stagnant thing in her lungs began to flow with it, up and out of her body. The plant a channel, as I had been a channel. And interwoven with that moving stream was the living teaching, the medicine understanding of this plant, this ally, this living being that people call skunk cabbage.

CHAPTER THIRTEEN

THE IMPORTANCE OF RIGOROUS SELF-EXAMINATION AND THE NECESSITY FOR MORAL DEVELOPMENT

We are well enough aware that some skill, some ability, usually predominates in the character of each human being. This leads necessarily to one-sided thinking. Since man knows the world only through himself, and thus has the naive arrogance to believe that the world is constructed by him and for his sake, it follows that he puts his special skills in the foreground, while seeking to reject those he lacks, to banish them from his own totality. As a correction, he needs to develop all the manifestations of human character—sensuality and reason, imagination and common sense—into a coherent whole, no matter which quality predominates in him. If he fails to do so, he will labor under painful limitations, without ever understanding why he has so many stubborn enemies, why he even meets himself as an enemy.

— GOETHE

[Thoreau] aimed to become just, and in this struggle followed the ancient doctrine, contrary to scientific doctrine, that certain secrets of nature reveal themselves only to the observer who is morally developed.

— ROBERT BLY

IN THE INITIAL STAGES OF DEVELOPING YOUR NATURAL CAPACITY for direct perception, the learning itself, the experience of the work, takes all the attention. But later, as with riding a bicycle, once you have experience with it you are not so concerned with keeping balance. The balance is, finally, automatic. You can begin to look around and enjoy the view. You know how to ride so well that you can do much more elegant things than merely going to the store. You begin to pay attention to riding in a different way. The riding itself, you find, gives you information about bicycles and balance and roads.

and yourself

There is a refinement. You begin to be aware of your heart-field as an entity in and of itself. You begin to know its shape, its identity. While you perceive through the heart and its field, you are not the heart itself, nor the field (just as you are not your brain or its field).

though the field in which you locate yourself
will shape how you perceive

Your heart-field has an identity, something that is more than the sum of the parts. It has a particular shape, even though that shape is in constant flux. It has a particular feeling, even though that feeling is in constant flux. It has a specificity that you can know as well as you know your hands.

By regularly using your heart as an organ of perception, you become sensitive to its shape and its quality and to every alteration that occurs in that shape and quality. Just perceiving in this way develops your heart as an organ of perception, brings its identity into your awareness. The reflection of the world within its field, and your attention to that reflection, lets you know yourself even more intimately.

The human being knows himself only insofar as he knows the world; he perceives the world only in himself, and himself only in the world. Every new object, clearly seen, opens up a new organ of perception in us.

— GOETHE

Because the field of the heart is so sensitive, everything that touches it causes it to shift its form. Its quality alters with each touch of the world. These alterations are ripples of waveforms—caused by living touches from the external world—traveling over the surface of and throughout the heart-field. The nature of these ripples, their intensity, height, duration, and form, contain—are—information about what is touching the heart. Just as the moon is reflected in the still water of a pool, these ripples are reflections from the world around us.

All the world goes by us and is reflected in our deeps.
— Henry David Thoreau

Thus, if you cultivate a sensitivity to being touched in this way, if you constantly sense through your heart, you will become aware of your heart-field as being very much like a still pool of water.

just as you know your hands

Anytime the heart-field is touched it is like a stone rippling into a still pool. Anytime the heart-field alters its shape, you can allow yourself to perceive it. In that moment of noticing the mind can be focused on the alterations in the field, and an image can be seen within it.

The heart-field

in some ways

is like a three-dimensional mirror, if you can conceive of such a thing. What you are doing now is something like working with a three-dimensional image reflected in a three-dimensional mirror. Once you know this, you can become attentive to it and actively notice anything that alters the field.

the notitia *of the ancient Greeks, the active noticing*

When you use imaginal seeing as a method of perception with the heart-field itself, your consciousness, your focused awareness, will become aware of any images that appear in the surface of this three-dimensional mirror.

We can never see directly what is true, i.e., identical with what is divine; we look at it only in reflection.
— Goethe

(However, to make this all more complex, the images that appear in this three-dimensional mirror do not simply appear on a flat planar surface but appear throughout it, within it, in depth. And, of course, the mirror is not three-dimensional but, like a mountain, possesses a dimensionality somewhere between two and three.)

The meanings from the world, from the particular thing you are perceiving, a plant for example, alter the fabric of this field. The particular alteration itself is a reflection, an image of the thing touching it.

Through your focused awareness, your directed consciousness, you notice how the field changes, and, with practice, can see the image it carries from what has touched it. You look through the surface of the mirror to see the thing, but the thing you are perceiving is not the mirror, not the field within which it appears.

The material substance of the mirror metal or mineral, is not the substance of the image. It is simply "the place of its appearance."
— Henri Corbin

The appearance of the image in the field of your heart is nearly instantaneous.

the speed of light is very fast

The alterations in the field are sent to the brain in a flash-burst of communication through the direct connections that exist between the heart and brain. The brain compares the alterations to the normal field identity, extracts the alterations in the waveform, and begins analyzing them for their embedded information. The meanings, the communications, from the external phenomenon that initiated the shift are understood in the same way that a plant or disease is understood.

Normally this begins when you sense something, feel something,

a different feeling than is usual

and focus on the feeling. This directs your attention to the field itself. You enhance the feeling, slightly disengage, wait a moment, then reengage. Eventually there will come a pregnant moment in which the meanings underlying the alterations in the field reveal their nature, when you understand the thing itself. In that moment, that pause in time, you can also determine from which direction the alterations in your field occurred, just which phenomenon touched you.

with experience this only takes a few seconds;
in the beginning it takes awhile

At this pregnant moment, as with all pregnant moments, depth knowledge of the phenomenon expresses itself through a unique mode of representation. And athough you have directly experienced the thing itself, it will take its form, its analytically understandable shape, from your preexisting store of feelings, experiences, memories, ideas, thoughts, and learnings.

The shape the phenomenon takes—just as with plants and organ systems—must be allowed to emerge of itself in whatever form it wishes. This means that you must be open to whatever mode of representation occurs.

> *[It is crucial] to cultivate as many modes of representation as possible or better, to cultivate the mode of representation that the phenomena themselves demand.*
>
> — GOETHE

You must remain as open as possible and let your seeing be shaped by the phenomenon itself. This requires a tremendous flexibility in your internal world. In consequence, this kind of direct perception initiates an unavoidable encounter with your own personal *history*. If you are doing depth diagnosis, for example, on someone who rather unpleasantly reminds you of the energy or mood put forth by your alcoholic father, you will be unable to see this new person in his own light—unless you are emotionally unattached to your historical experiences with your father. Your analytical mind will generate a mode of representation similar to that of your father, and while this is informative,

> *Your father's face might flash in your mind's eye,*
> *you might then feel the feelings you associate with him.*

the unfinished emotional baggage that accompanies this representation will interfere with your ability able to clearly see the person in front of you.

> *maturity, rather pedantically*
> *is no longer being able to fool yourself*

The flash of your father's face before your inner sight is information about the person in front of you. But . . . this person is not your father; you are not his child. And your father is not now engaging with you. However, the mode of representation that has occurred is a representative gestalt that tells you something essential about the person you now see.

And you must now decipher this mode of representation in order to understand the living reality in front of you. If you get distracted from the task before you, reengage with unresolved issues you have with your father,

> begin thinking about all those things
> you have thought about before,
> start down the well-trodden trail once more,

or you will never see the person in front of you with transparent eyes. The mode of representation that arises is about the person in front of you. What you do with it depends on you.

All human beings possess these unclarities—these histories. They are an inevitable aspect of the human predicament. A dedication to this mode of perception, however, forces personal transformation in order for you to master the process. Undifferentiated application of these old memories and unmet needs is what is often referred to as projection.

> Seeing the world exclusively through the analytical mode is another form of projection—mechanomorphism.

You must have a drive to see with transparent eye, to have no judgments about, desires for, or emotional aversions to the mode of representation that arises within you. This calls for tremendous personal awareness.

> All mean egotism vanishes. I become a transparent eyeball.
>
> — RALPH WALDO EMERSON

Unresolved emotional experiences, when activated, can move up from unconsciousness and affect our sight. Like a film across the surface of the eye, these experiences distort, alter, and shape what we see.

> The least partialness is your own defect of sight and cheapens the experience fatally.
>
> — HENRY DAVID THOREAU

Thus when working with feeling as a sensory medium, you will have problems if you do not understand that activation of old emotional experiences is an integral part of the process.

> The manifestation of a phenomenon is not detached from the observer—it is caught up and entangled in his individuality.
>
> — GOETHE

Personal histories are simply old experiences upon which the brain draws when assembling a gestalt of understanding. If you do not understand them and have not come to some sort of resolution with them, you will perceive them as real when they are simply a mode of representation.

Whether or not they are completely resolved is unimportant; it is our self-awareness of them that is crucial—our knowledge of how they generally affect us and our ability to step aside, to not reengage with them, while still taking them into account as information and remaining genuinely open and emotionally connected to the person in front of us.

The continual practice of this mode of cognition, the process of engaging and then disengaging with the phenomenon, precipitates psychological unclarities, moves them forward into conscious experience. For the more you do this, the more modes of representation you will experience.

> *Above all, true researchers must observe themselves and see to it that their organs remain plastic, and also remain plastic in their way of seeing. So that one does not always rudely insist on one mode of explanation, but rather in each case knows how to select the most appropriate, that which is most analogous to the point of view and the contemplation.*
>
> — GOETHE

The intention to grow—to master the process—so that you can see with a transparent eye forces the removal of preconceived patterns and meanings from the work and allows the organizational pattern of complex interrelating to directly impact the perceiving mind and heart, and enables the meaning itself to emerge into perceptive understanding *in its own mode of representation.*

In essence, this means that you understand your heart-field so well that you perceive it like the still surface of a pond. Any image that strikes it is allowed to pass directly within you, to be held within you, to take a shape with which you do not interfere. You remain a detached observer, but in a unique sense. You are feeling. Not thinking in a Sherlock Holmes, Mr. Spock, Data, (Hannibal Lector), sort of way, but rather are tightly interwoven with the phenomenon you are experiencing. The phenomenon is entangled in your individuality and you are entangled with its individuality. Deeply experiencing. Deeply feeling. You are not detached. But at the same time a part of you is observing, seeing the thing take shape before you in all its multidimensional complexity. As you feel into it, your entire field of perception changes.

The meanings in the thing flow through you: your emotional body changes, your feelings change, your perspective changes, yet your perceiving awareness remains unchanged. And this perceiving awareness is not your linear, thinking mind, but something else entirely. It is the consciousness that can flow to any biological oscillator in the body.

To grasp the phenomena, to fixate them into experiments, to order one's experiences, and to come to know all the ways in which one might view them; to be attentive as possible in the first case, as exact as possible in the second, to be as complete as possible in the third, and to remain many-sided enough in the fourth, requires that one work one's poor ego in a way I had else hardly thought possible.

— GOETHE

Thus, because you are so deeply perceiving meanings as a participatory consciousness, the process calls for a commitment to rigorous self-examination. You trust your senses, your heart's ability to perceive in this way, and yet simultaneously keep a healthy skepticism of your internal response to the mode of representation. The key word here is *response*.

The senses do not deceive, judgement deceives.

— GOETHE

The mode of representation will arise of itself. It is your response to it that you must examine closely.

In observing Nature on a scale large or small, I have always asked: who speaks here, the object or you?

— GOETHE

You must trust your internal responses, trust what your senses tell you, but realize that they are about the phenomenon that is outside you. These responses are not about you.

His eyes were tiny pinpoints, sharp-focused. Watching. His hair lifeless and washed out. His skin, too, had an odd color and texture, fish-belly white, orange-peel texture. And he was strong; the power of his stocky body filled the room.

As I listened to him talking, I had a sudden overwhelming desire to beat him up, hit him, hurt him, throw him out of the office. The image of my violence flashed on the screen of my vision. I could feel my fists hitting his flesh. And a part of me liked it.

I was shocked. Horrified. And began to focus on myself, trying to root out this violence inside me.

"Healers aren't supposed to feel this way," I thought.

Then my training kicked in and I stopped myself. I breathed deeply and let go of my emotional response to the image.

I kept talking to him, letting the image remain within me. Feeling into it. Seeking its source.

Then suddenly, into my awareness, came his smell. A smell that existed below the level of a noticeable odor. And I realized that at some deep level he smelled bad. It was the same kind of smell, I realized, that causes birds to throw their babies out of the nest. A smell from a sickness deep within the organism. A wrongness that could not be denied.

Immediately I asked him, "Was there anyone you were close to growing up? Do you have any really close friends or family now?" I was not surprised when he said, "No." Nor was I surprised later when I saw him in a room of people and noticed them unconsciously flowing to the other side of the room.

You must gain a facility with understanding the responses you have to the modes of representation that arise within you, and learn to untangle their meaning.

Sometimes, when you first begin to learn this work, the feelings you get from an outside source can become tangled up in your internal responses, as if a pile of different colored threads were dumped in a big heap on a table. To sort them out, you must take each thread as you find it, slowly pull it out of the pile, and follow it to its end. (Then follow it back again to the other end, back and forth, until you know it well.) In this way you will know which threads of emotion/meaning come from outside you and which are response/meaning threads from within you.

Each thread that you pull from the pile must become a contemplation, a meditation. You must allow yourself to drop down into each one, to come to understand it as clearly as you did the plants. You must allow them to unconceal themselves, as all things will if you consider them in this way. In time, you will come to know each response thread that you have, and these you will bundle up, each on its own, all together in a basket you keep inside you. You will know each in terms of its own unconcealed identity.

Then, when you are working to decipher meanings and sense some confusion, you can seek the response within you that is responsible. You ask, send out a request from the self, and from the basket, from the database of knowledge that you have crafted, the identity that you seek will emerge. It will rise up before your internal vision on its own, just as the plants do.

You will remember that identity then, know its thread of meaning, experientially touch again its axes of dimensionality. You can then reach into the meanings you are working with and, slowly, untangle your responses in order for the phenomenon to stand before you in its own light. Each of these threads represent something about you that you do not want to see, and they will confuse your sight if you do not come to intimately understand them. Each thread leads deep inside us, to a place in which we keep unexamined parts of ourselves.

Thus, obtaining clarity in this process necessitates a personal engagement with your hidden selves, with your personal demons, and with your shadow. An engagement with the many different parts of you, the feelings you have, your history,

the parts of you that you have sundered from yourself

aspects of yourself you do not want to see. For one thing is true: looking into this mirror to see something gives you back your own reflection as well. This mirror that is the field of the heart reflects everything. It reflects not only the deep meanings of phenomena that you encounter, but also a unique secondary reflection.

the mirror within the mirror

The secondary reflection is the distortion of the phenomenon that comes from your internal unclarities—your responses. The kind of distortion that occurs is itself a mirror of your internal world, of the things you do not want to see. To see clearly, to "become a transparent eyeball," you must learn to look below the surface of the reflection in your heart-field, to see the secondary reflection that comes from your deep self. And then to begin the work needed to come to terms with those hidden aspects of yourself, to heal them, to reincorporate them, to reintegrate the split off parts of yourself back into your Self.

The mirror in your bathroom reflects your physical image. This mirror reflects everything that is within your image. Seeing that interior reflection means encountering the uncomfortable. This reflected truth is something few want to see.

Who was it that injured you?
Were you very young?
You must have been, for
I see your mouth moving
seeking a nipple,
or some sustenance from life,
that a deeper part of you
has longed for,
but has never been able to find.

I wake in the night sometimes
and find you curled
in the shadow of my arm.
Feet drawn up, thumb sucking
and it takes an effort
to pull the wrinkled member
from my mouth
to relax my legs
and straighten them out

under the cold,
emptiness
of the sheets.

But seeing it is essential. Unlocking its mysteries and doing the work that is indicated is imperative for a successful mastery of this mode of cognition. To attain mastery of this work, you must be willing to see—and reincorporate—the shadow parts of yourself that you have hidden away.

When we were one or two years old we had what we might visualize
as a 360-degree personality. Energy radiated out from all parts of our

body and all parts of our psyche. A child running is a living globe of energy. We had a ball of energy, all right; but one day we noticed that our parents didn't like certain parts of that ball.

— Robert Bly

These shadow parts, the parts that others didn't like, are perhaps hidden away in a little room with the key cleverly concealed or maybe, as Robert Bly says, we have stuffed them into a long bag we trail behind us.

Into the bag, from the beginning, we put the parts of ourselves that we learn are unacceptable to our parents. One day they turn to us and say,

"Be quiet!"

And so the loud part of us goes into the bag; we learn to keep ourselves quiet.

"If you loved me you wouldn't do that."

And the part that we now think kills love is put in the bag.

"It isn't nice to try to kill your sister."

And now the predator part of us goes into the bag.

which is, of course, a very dangerous thing to do, and a very dangerous place to put it

By the time we are twenty, the bag is a mile long—all the parts of ourselves that we do not like or think are bad or wrong or unlovable or useless are in there. And the process of stuffing the bag has become automatic. The many parts of ourselves that are in the bag are no longer in our conscious awareness; they are now part of our unconscious. Perhaps by the time we are twenty only a tiny sliver of us is not in the bag—a sliver that exists just above the eyebrows and about two inches into the skull. We have reduced ourselves, become linear,

one-dimensional

convinced that we are only a single point of view, believing that multiple

multidimensional

personality is the exception, rather than the norm. The depth of us is in a bag, a bag we no longer remember, filled with parts that we no longer know we have.

Behind us we have an invisible bag and the part[s] of us our parents
don't like, we, to keep our parents' love, put in the bag. By the time
we go to school our bag is quite large. Then our teachers have their
say . . . [And] then we do a lot of bag-stuffing in high school. This
time it's no longer the evil grownups that pressure us, but people our
own age.

— ROBERT BLY

But those parts of us are essential to our wholeness. They are part of
us. They can never be removed, only hidden away.

> *perhaps in a bag*
> *or in a little room,*
> *the door (securely) bolted,*
> *the key hidden in a secret place*

Eventually, the realization dawns on each of us that these parts of ourselves, since their imprisonment, have had little to do except plot escape.
(No one likes being locked up in a bag or a little room, you know.) It isn't
very long before they find the key we thought so well hidden. Then they
begin to come out, when our minds are someplace else, occupied with
some other thing. And they are not amused.

> *Was it your love that put this here*
> *or the small secret part of you*
> *that you hide in darkness?*
>
> *Both smile enticingly,*
> *hands spontaneously lifted*
> *to touch my cheeks.*
>
> *But I have noticed,*
> *that that one's teeth*
> *are slightly longer.*
>
> *And she looks slyly*
> *out of the corner of your eye*
> *as she takes the sharpened words,*

and slides them home
in the part of me
that love has made defenseless.

For any part of us that we put in the bag

as with all prisoners

becomes distorted, unhealthy, insane. Life is not meant to be lived in a bag. Moral conscience is lost—the part just wants out.

Most of us are unaware of this process, of course. Only later in life do we discover our error, begin to understand that something is wrong, that these split off parts are essential to our wholeness.

that our drive to be normal is itself terribly abnormal

And so we try to get our parts out of the bag. But the problem is that we no longer remember how many parts there are, or exactly where in the bag we put them. Sometimes we no longer remember that there is a bag at all.

We spend our life until we're twenty deciding what parts of ourself to put into the bag, and we spend the rest of our life trying to get them out again.

— ROBERT BLY

And what's more, those parts are enraged. So when we catch a glimpse of one with our conscious mind, it scares the begeesus out of us. For another truth that comes in its own time is that any part that we have locked up begins to take on tremendous energy. And the years of repression, the stored energy of being

shut up!

have taken their toll. The shut-up parts have begun to devolve, to become hairy and monstrous, to grow claws.

solitary confinement always does this

We begin to notice these put-away parts, reflected in behaviors to which we paid no attention until a lover or boss or close friend calls them into our awareness. A light comes on and, unexpectedly, the part throws a shadow we now can see.

We notice that when sunlight hits the body, the body turns bright, but it throws a shadow, which is dark. The brighter the light, the darker the shadow. Each of us has some part of our personality that is hidden from us. Parents, and teachers in general, urge us to develop the light side of the personality—move into well-lit subjects such as mathematics and geometry—and to become successful.

— ROBERT BLY

Perhaps our conscious part, our successful part, is a healer who we've trained to be good and kind and filled with light.

Dr. Jekyll

but the rest of him, put in the bag, is dark, apelike, barbaric, and cruel.

Mr. Hyde

A dynamic part, powerful and filled with the energy of years of repression.

The nice side of the personality may be a liberal doctor, for example, always thinking about the good of others. Morally and ethically he is wonderful. But the substance in the bag takes on a personality of its own; it can't be ignored.

— ROBERT BLY

And when you finally accept that this part really exists and discover that you can't get rid of it by taking a pill

another kind of bag

you realize that you must encounter it directly. So you begin to grapple with this part, and you find that it is elusive and distrustful of your intentions. It has no desire to be forced back into the bag. (And of course, that is always your original intention.) You find that the part likes being out. It likes expressing its rage in clever ways.

sorry about the lace tablecloth and that glass of wine

But over time, as you cannot get rid of it, you resign yourself to its reality. You begin to watch for it, eventually to engage it in conversation. And, realizing that it is a part of yourself, integral to the wholeness you desire, you begin the struggle to make relationship with it, to come to care for it, to reincorporate it into yourself. If you keep on, you discover that while these parts can no longer be commanded, they do understand bargains. For there are things they want, things only you can give them. And there are things that you want, too, you find in time, that only they can give.

Gradually, as you become trustworthy to yourself, that part will begin to trust you again, will eventually reintegrate with you, and your sphere of energy will expand from a sliver, becoming a bit larger with each new part of you that you recover, establish relationship with, keep your word to, and come to love again.

You find, too, as you interact with the part, that as its rage lessens, its shape begins to alter. And one day you look at it and notice that it is no longer hairy, its teeth no longer long and sharp, it fingernails no longer clawlike. Its face looks somewhat like yours and you notice he is a boy of a certain age, with certain skills and talents,

certain points of view

that you need in order to perceive the world from a 360-degree perspective.

Beginning this reclamation of the self

this soul retrieval

means stopping one day, turning, and seeing the long bag that you trail behind you. It means opening the bag, reaching in, taking out some shadow and eating it.

Eating shadow conveys a certain moral authority.
Others instinctively notice
that you are not afraid of the dark,
that you possess a depth of self,
that you have eaten something
they have not.

Eating shadow, eating the interior shut-up parts, stops your unconscious need to eat external interiors. Stops the (linear) progression from Sherlock Holmes, to Spock, to Data, to Hannibal Lector, the last of whom can only survive by eating the interiors of others.

the linear mind eating the heart of the world

When you first open the bag and let a part out, it is often misshapen, frightening, and angry. It takes some time, much bargaining, many kept

promises, and much love before it once again takes on its proper form—
before it is healed and whole.

> *and these parts must genuinely be liked*
> *a genuine relationship reestablished*
> *for any reintegration to occur*

It takes a certain strength of character to reabsorb your own shadow,
necessitates an inevitable encounter with the darkness within you. Any
part, locked in its room, stuffed in a bag and never allowed to roam free,
becomes more and more damaged as time goes by.

Reclaiming all the parts of the self locked away is an essential part of
moral development. It is ecological reclamation of the self.

> *Let yourself become comfortable in some place where you will not be*
> *disturbed. Breathe deeply and let the tension go out of your body. Let*
> *yourself be held by the place in which you are sitting. Let yourself relax.*
>
> *Now. See standing in front of you the ugliest part of your body.*
>
> *How does it look? How do you feel seeing it? Is there anything*
> *in particular that it wants from you?*
>
> *Are you willing to give it what it wants?*
>
> *Feel how much power and energy that part of you has, how*
> *much energy and power it takes from you every day to keep it locked*
> *up inside. What would you do if you had all this energy to use every*
> *day for something else?*
>
> *What would it take for the two of you to become friends?*

This mode of cognition you are developing, your continual perception
of meaning, stimulates of itself this internal ecological reclamation.

> *The wounds in the world*
> *are reflected within us.*
> *Or . . . did those outer wounds*
> *begin long ago*
> *when we locked a part of ourselves*
> *away?*
> *Who can say just where*
> *the reclamation of the world*
> *begins?*

For when we enter the wildness of the world, begin perceiving with the heart, reading the text of the world, an energy from the world flows deep inside us.

It flows up through you and out again, back into the world. Any place within you that is bent or twisted casts a shadow. A boss or a lover or a close friend may tell you about a part locked away, shine a light that throws the shadow of that part on a wall so that you can see its shape. But the world shines a light that reveals everything by the shadows that it casts. And these shadows, their shape and orientation, let you know in just what ways you are not straight within yourself. The force of that stream of meanings pushes against the bend and you feel a pressure within you, a pressure that is a movement towards straightness, towards becoming upright. In the light of the world, every part of you that is in the bag will eventually be revealed. And along with that light comes not only revelation, but also the teachings necessary for personal ecological reclamation to occur.

There is death in this restructuring, this reclamation of the self. For the reduced-you dies when you eat shadow, adding a new part to yourself.

it is hard work

This mode of cognition ultimately changes everyone who engages in it. To be true to the process, to master this way of seeing, means that there can be no nook or cranny of yourself that does not open itself to your gaze.

or to the gaze of the world

It means that there can be no part of yourself that you do not come to know, that you do not release from its locked room and reintegrate into your whole self.

> I have seen myself lose intolerance, narrowness, bigotry, compla-
> cence, pride, and a whole bushel-basket of other intellectual vices
> through my contact with Nature . . . The raw materials to make this
> growth possible, though, do not come from introspection or selfish-
> ness. They come from the application of lessons from without—the
> influence of environment, repeating itself over and over again on the
> sensitive plate of the brain and being transformed there, as sunshine
> and floods are transformed in the leaf of the plant, into material for
> the beautifying of the mind and the enriching of the soul. You have
> it in your power only to keep the brain sensitive to impressions and
> the heart and mind and soul adaptable to growth; you have "the

power to vary," and the extent to which you utilize and benefit from that power depends on you and no one else.

— LUTHER BURBANK

The process is a long one. Each year that you use this mode of cognition the more phenomena you will encounter, each of which will demand a greater clarity from you.

But if the observer is called upon to apply this keen power of judgement to exploring the hidden relationships in Nature, if he is to find his own way in a world where he is seemingly alone, if he is to avoid hasty conclusions and keep a steady eye on the goal while noting every helpful or harmful circumstance along the way, if he must be his own sharpest critic where no one else can test his work with ease, if he must question himself continually even when most enthusiastic— it is easy to see how harsh these demands are and how little hope there is of seeing them fully satisfied in ourselves and others. Yet . . . this must not deter us from doing what we can.

— GOETHE

And just when you think you have gone as deep as you can, when there are no more shadows to find, no more parts locked away, a phenomenon will thrust itself upon you—a person, place, or plant—that forces you to go deeper still. Forces you to see parts of yourself that you did not know were there, that you had no

conscious

wish to know.

You will find that once you begin this process, the world will shape you; the things that come to you will be the ones you needed to meet in order to become yourself, to attain a 360-degree perspective once more. You will find, as others have before you, that this mode of cognition is a soul-making process, that indeed, as the poet John Keats said, "The world is the place of soul making."

And, of necessity, you will find that a moral development occurs within you. The shaping that is happening to you, the continual demands that you know yourself, that you face the darkness within and without, begins to take on a moral dimension. Not in the inadequate, religious sense of that mistreated term, but in its original sense, meaning your shape, your structure, and how you are arranged internally.

being upright

And this rearrangement of your internal self is stimulated and directed from the world outside you. It does not come from a top-down hierarchy of values, but is a quality, a value, that emerges from within, from the center outwards, when your individuality is entangled with that of the world. It comes out of a center of expressed meanings, out of interaction with the world, out of aisthesis.

The field of the heart contains within it everything you are, everything that is within you. Every thought you have, every unmet desire, psychological need, and wound. Most of these things are unconscious. It is in meeting them in the mirror, in the reflections given back to you from the phenomena of the world, that you see them and begin to make peace with them, to know them, to work them through, to reintegrate them into yourself.

> *The mirror itself changes,*
> *its very substance is altered*
> *by the process of you becoming yourself.*
> *Its capacity for undistorted reflection*
> *begins to improve.*

The imposition of a top-down morality, the forcing of a static, linear morality onto the self, does not alter the essential self, only buries it deeper, under more layers of oppression. A top-down morality insists these layers must not be pulled back, that what is hidden under them never be revealed.

> *The deepest evil in the totalitarian system is precisely that which makes it work: its programmed, single-minded monotonous efficiency; bureaucratic formalism, the dulling daily service, standard, boring, letter-perfect, generalities, uniform. No thought and no responsiveness. Eichmann. Form without anima becomes formalism. . . forms without luster, without the presence of body.*
>
> — JAMES HILLMAN

It is perhaps why many who possess a top-down morality never want to look into the world and truly see what is there. For the reflections they receive would, indeed, be terrible to bear.

But the morality that comes from engaging with the world is not a top-down morality. It is something else again, a living thing that comes from the world itself.

You are shaped by this interaction with the world in the way you are meant to be shaped. In your nonlinear, unprogrammed walking through the world, you find the things you need to find. The clients who come to you, the plants you meet, the particular wildness of the world that you eat, or that motivate you to eat any particular shadow from your bag, all shape you in particular ways. A certain kind of morality begins to emerge. You begin to take on the luminous quality present in healthy ecosystems, in old-growth forest, in mountains.

> *Once we have . . . "fallen in love outwards," once we have experienced the fierce joy of life that attends extending our identity into nature, once we realize that the nature within and the nature without are continuous, then we too may share and manifest the exquisite beauty and effortless grace associated with the natural world.*
>
> — JOHN SEED

True morality begins to emerge of itself. The continual exchange of soul essence, of heart-field, of communications with the wildness of the world allows the wildness of the world, and its essential morality, to enter within.

> *Preconceived notions, dogmas, and all personal prejudice and bias must be laid aside. Listen patiently, quietly, and reverently to the lessons, one by one, which Mother Nature has to teach, shedding light on that which was before a mystery, so that all who will, may see and know. She conveys her truths only to those who are passive and receptive. Accepting these truths as suggested, wherever they may lead, then we have the whole universe in harmony with us. At last man has found . . . that he is part of a universe which is eternally unstable in form, eternally mutable in substance.*
>
> — LUTHER BURBANK

Because the phenomena upon which we focus our attention penetrate so deeply within us, we are deeply touched by the meanings that they embody. These meanings themselves have tremendous impacts on how we perceive ourselves and the living world within which we are embedded. The human organism naturally restructures itself around the meanings that are experienced. This forces us to do internal work to allow that restructuring to happen without twists, without bends. Thus encountered meaning reorients the human, entrains the human, to reflect that meaning.

Because Nature does not lie, the direct perception of Nature means that each of us who does lie, each part of us that lies, even in our deep unconscious, must reorder, must restructure, if we truly want to perceive deeply into Nature.

> *Each part of us in the bag of shadows is a lie,*
> *an internalized lie from our teachers or peers,*
> *or a lie from ourselves*
> *to ourselves.*

The more we lie, are out of accord with the truth that is found in Nature, the less we are able to perceive of the depth dimensions of Nature. The hidden face of Nature, thus, is an expression of its moral dimensions, which are as real as its physical dimensions. We partake of the moral not because we are human, but because we are of Nature.

> *It is by obeying the suggestions of a higher light within you that you escape from yourself and, in transit, as it were see with the unworn sides of your eye, travel totally new paths.*
> — Henry David Thoreau

As the work continues, there is an alignment of unconscious motivations, fears, and drives with Nature. As this reorientation occurs, you will begin to perceive deeper communications from the world. The teachings will deepen. For each thing rightly seen unlocks a new faculty of soul.

> *The Sal is what was there all along, the part you can't see until you get rid of everything else.*
> *It's your birthright.*
> — Dale Pendell

The *intensity* of the personal struggle most of us go through in developing this kind of morality is, in many respects, a reflection of our too-long primary use of an inappropriate mode of cognition, the linear, analytical mode. The verbal/intellectual/analytical mode of cognition is, by nature, amoral. It is also exceptionally shallow.

The insistent, single point of view of the linear mind is itself a lie. All people are in fact born multiple personalities. All people naturally should possess multiple points of view, have a multidimensional consciousness. The adoption of a linear, *single-minded* focus corrupts the self, forces us to forgo depth of self and to become one-dimensional. It forces parts into the bag by its very nature.

The holistic/intuitive/depth mode of cognition is inherently multidimensional, deep, and nonlinear. You cannot help but be changed if you engage in it. Engaging in it, in and of itself, thrusts each of us into the personal experience of a multidimensional reality, in the face of which the linear mind cannot maintain its single-pointed focus.

> *Nature is both the creator of man and his greatest teacher. Sensitivity, reason, and understanding true to man all can be manifested only through sympathy with nature. Judgement and criteria for right and wrong, virtue and evil, excellence and mediocrity, beauty and ugliness, love and hate do not hold if man steps off the Great Way pointed out by nature.*
>
> — MASANOBU FUKUOKA

As we find, then reclaim, the parts of ourselves that we put in the bag, we establish relationship with them and learn that we must keep our word to them, become honorable again.

Integrity: the state of being whole and undivided.

Breaking one's word creates a division between self and other, the contract that the word sealed rent asunder. Our (formerly) shadow parts must come to experience, through our behavior, that we can and will keep our word, no matter how difficult our personal situation becomes.

a person's word is often only as good as her comfort level

It is only in difficult times, when personal discomfort is extreme, that the power of our word to ourselves is tested. Passing through these points of discomfort with word intact shows more than any talking we can ever do that our word is indeed good. And this is essential training, for Nature must also know that we can keep our word. The plants do not understand about human social concerns, about soccer practice, or being too busy. Nature will, at some point in the process, need to know if you are serious. Serious enough to remake yourself in her image, to allow the image, the Nature that you are, to emerge from within you. Serious enough to be true.

> *The forest person will scrutinize you closely. Maybe she will sniff you. Maybe not. But she will size you up, that's for sure. Sacha Huarmi wants to see if you are complete, if you have the fiber to see the training through to the end.*
>
> — DALE PENDELL

Are you truly willing to do what it takes to master this? Are you willing to enter the fruitful darkness and find the fertilizer that lies at your feet, around your roots? Are you truly willing to see with transparent eyeball? If so, you will begin to eat shadow and remake yourself out of the teachings that come from the world.

Once you do, there will come a time when the work deepens of itself, when suddenly you begin to see in ways you had not thought possible.

> *Nature is an exacting mistress and a jealous teacher; she does not reveal herself wholly to the amateur or the dabbler, and she will not cooperate fully and generously with the man who takes her lessons or her work lightly.*
>
> — LUTHER BURBANK

You suddenly find one day that you have become the servant of Nature. That you have eaten the wild so long that you are changed. And this change

and the work that you have done within yourself

has caused you to regain a moral stature, in accordance with the deep truths that permeate all Nature. You will understand in your experience that She is the source of all things, that from Her all life has come and to Her all life returns.

And in this service to Nature you surrender the dominance of the linear mind, understanding, finally, that surrender does not mean defeat, but life itself.

Nature understands no jesting; she is always true, always serious, always severe; she is always right, and the errors and faults are always those of man. The man incapable of appreciating her she despises; and only to the apt, the pure, and the true, does she resign herself and reveal her secrets.

— GOETHE

And you enter, completely, into the world of Nature. No longer having one foot in this world and another in that. You give up being a bridge and cross over. You enter into a conversation—an interblending of soul essence—that is deep and old and basic to our species. And what you find there has been spoken of by ecstatic poets in all times and places. For you begin, when you make this transition, to read every day the book of Nature and to hear the sacred as it speaks to you from all created things.

God is the creator of all Nature, but this Great Spirit may also be thought of as lying hidden within Mother Nature as the force that raises and nurtures it. The form of God finds expression in the form of Mother Nature; mental images of the heart of God may be thought of as arising from within Nature and being caught by man. Thus, the breath of God becomes Nature, and the heart of Nature makes man human.

— MASANOBU FUKUOKA

GRAINS OF SAND FROM ANOTHER SHORE

*There was a time
when I saw the world
Coyote lives in.*

*I had walked up,
with a friend—once upon a time,
behind the rocks,
the big ones that rise up, mossy-greened,
and cradle the forest-shadowed ponds
that the ducks and moose love,
to seek the slight-sloping, grassy meadow
hidden behind them.*

*We half-lay for hours
in the tall emerald grass
among the ancient trees that towered over
the drifting textures of the land.
While our elbows supported us
we talked of plants,
and stones,
and the wisdom of moss.*

Slowly we began,
as humans sometimes do,
to slip into the wildness of the world.
Our language began to slow
down, pause, and falter.
Into silence we drifted
and for some reason
that only our souls understood that day
we flowed with it, not talking.

Colors became more vivid
and the air began to sparkle.
Our breathing and the sounds of forest
took on a luminous quality.
And into this silence Coyote ambled,
following a game trail
that flowed, brown runnel, near our feet.

Her tongue lolled out
the side of her mouth,
and she was laughing
that crazy laugh Coyote has,
while her eyes spun
as she watched the dancing bones
that lie under the fabric of the world.

Crazy, gamboling Coyote.
Third force in Universe.
I said under my breath,
"Turn your head to the right."
And my friend sat up
and said, "What?"
And in so doing, lost her chance to see.

I, still watching, saw Coyote's eyes
shift out of that crazy, spinning universe
and shocked,
no,
betrayed,

by the secrecy of our immersion
she flipped straight up and over
and ran, tail between her legs,
only some strange kind of dog,
up the trail.

What I glimpsed through Coyote's eyes
lodged in a part of my brain
I did not know I had.
I can reach out and touch it sometimes.
My eyes begin to spin,
and I feel a bit dizzy,
and I can see
dancing bones
under the fabric of the world.

I still do not know
what the world
that Coyote lives in
does
when no one is watching
but I do know it is ancient
far beyond the species lifetime of humans
and that next to it, our world
is only a chip of wood
floating on the ocean.

CHAPTER FOURTEEN

READING THE TEXT OF THE WORLD

THE GEOGRAPHY OF MEANING AND THE MAKING OF THE SOUL

As soon as you begin to read the great and loving God out of all forms of existence He has created, both animate and inanimate, then you will be able to Converse with Him, anywhere, everywhere, and at all times. Oh what a fullness of joy will come to you.

— GEORGE WASHINGTON CARVER

Being that can be understood is language.

— HANS-GEORG GADAMER

[Nature is] the Manuscript of God.

— LUTHER BURBANK

The World is the place of soul making.

— JOHN KEATS

THERE ACTUALLY ARE PEOPLE WHO HAVE NO ABILITY TO PERCEIVE meaning in even the simplest of things. They cannot tell that a table is a table, or a lamp a lamp. They are trapped in direct sensory experience. They receive sensory inputs, but cannot turn the sensory impressions into meaning. Even to recognize that a book is a book is a process of perceiving meaning. We live in a world, not of sensory objects, but of meanings. And those meanings are inherent in the things themselves.

In making a book, human beings insert the meaning into the physical object. But in the wildness of the world, the meaning was inserted long before books and printing presses or even human beings existed.

> *There is some essential identity, a meaning in plants themselves that allows us to recognize as plants those that are not remotely similar in form, shape, color, or environment.*

Our perception of the meanings in the phenomena around us connects us to those meanings; observer and observed become linked through the process of perception.

> *The error of empiricism rests on the fact that what it takes to be material objects are condensations of meaning. When we see a chair, for example, we are seeing a condensed meaning and not simply a physical body. Since meanings are not objects of sensory perception, seeing a chair is not the sensory experience we imagine it to be.*
>
> — HENRI BORTOFT

Because the condensation of meaning occurs almost immediately when we see something physical, we do not notice the process happening. It is so automatic that we miss the fact that we are doing it. But for babies, this is not an automatic process. When they see something for the first time, they do not ascribe meaning to it. They experience the meaning directly.

> *which is one of the reasons*
> *they are such a delight to be with*

Their inter-being with the world is intact; their experience of the thing and the thing itself flow together in a constant exchange of meaning.

To reclaim this ability does not mean we must go back to being babies, but it does mean reclaiming the capacity for direct perception that we once knew as babies,

and which is our birthright

a capacity that (and the memory of which) is still within us.

The automatic condensation of meaning that occurs beneath our conscious awareness can easily be perceived when we look at a sentence on paper that is written in the language we normally use.

"How are you doing?"

Such a sentence is immediately understandable to a reader, the meaning automatically attributed as it is read. But the same sentence in a language with which we are unfamiliar

"Wie geht es ihnen?"

is not,

unless you understand German

even if both say the same thing. With the first example, you automatically condensed meaning as you read. The second example is merely marks on a page, different in form but not in sensory impact from the giraffe figure before you understood it.

As with the giraffe figure, the meaning in the sentence, "How are you doing?" is not *in* the sentence itself. It resides in how the parts combine, the tension between those parts, which parts are next to the other parts.

this is true of each of the words and their letters as well

When someone is learning to read, they puzzle over the sentence just as you, perhaps, puzzled over the giraffe figure. Eventually, the meaning bursts upon them, just as the giraffe emerged out of the patchwork all at once.

In the real world, the meanings are expressed out of living organisms, not dead and static words on a page. Meanings are always perceivable by human beings. If they were not, we could not understand language, and you would not be reading this book. It is our capacity for perceiving meaning that is primary, not language. Language came second, out of our capacity to perceive meaning. Our language is a created form expressed out of the original nonverbal languages that human beings have always apprehended. It is a shadow, a reflection, a copy. Human language is only a special instance of language.

for in the beginning was language

Each phenomenon that we experience is its own language. It speaks always whether or not we are listening, for we are not the primary objects of its affections. All phenomenon, all plants, give off meanings that alter their structure—their content—in a continual ebb and flow. We are meant to perceive them with a mode of consciousness other than the linear, analytical mind. This other mode is accessed through the heart, as it feels the impact on us of the meanings we encounter. The analytical mind, used as a subsidiary system, a support to the primary system of perception, can translate the communications and feelings we experience into bursts of understanding that include usable analytical language.

> *Whilst there can be meaning that is non-verbal, there cannot be meaning that is non-linguistic for much the same reason that there cannot be a triangle that is not three-sided.*
> — HENRI BORTOFT

We are engaged in communicating through a highly complex, nonverbal form of linguistics, of which our language is only a reflection. Our brains perform an act of translation. This puts salt on the tail of the experience, lets us use it in our daily lives. But this translation must be continually reconnected to the origin itself, otherwise it risks becoming a dead thing, like much of science. It must continually be renewed by going into the world and directly encountering its images, over and over and over again.

Euclid took a system he created with the intellectual mind and applied it to the world. He did not let the world speak in its own terms; Euclid refused to see the nonlinearity of Nature. Descartes, through his analytical projections, created a separation between me, here, and the world, out there. He made an assumption that there is an external world of sense objects that exists independent of any observer. He formalized a mind–body, subject–object dualism, one that exists inherently *only* in the analytical mode of consciousness itself. It is the mode of perception that creates Descartes's world, it is not present in the world itself.

> *The Cartesian dualism and the onlooker consciousness are psychological consequences of emphasizing the verbal-analytical activity of the mind. Descartes's philosophy is therefore a projection of the psychological state which he produced in himself.*
> — HENRI BORTOFT

Within the images that the technological world offers us, there are no living meanings, only the appearance of meanings. Scientific technology takes the image, reduces it into something that is not real, but still has the appearance of reality. We lose the imaginal for the merely imaginary.

television

We must continually reexperience the living images of the world, the plants, and all of creation or we ourselves risk becoming a pale reflection of life, a copy, a shadow.

Direct depth perception of a plant or any phenomenon in Nature will always reveal dimensions to its being that science can never see because those dimensions are invisible to the linear mode of consciousness.

> *The recognition that the objects of cognitive perception are meanings and not sense data, shows us that "the world" is not an object, or a set of objects, but a text. Cognitive perception is not simply sense perception, in which material objects are encountered through "the window of the senses." It is literally, and not metaphorically, reading the text of the world.*
>
> — HENRI BORTOFT

Directly encountering the living plant through this mode of cognition allows you to experience the meaning of its text without an intermediary. Like understanding this sentence without having to read all the words. You are literally "reading the text of the world," the text of the plants that you meet individually when you travel into the wildness of the world.

> *ah, this is what you mean when you call it "doing readings."*

But this text that we are reading, it is more than mere words on a page.

> *The use of the word "text" falsely indicates an observer and an observed. It does not express the participatory interweaving involved in this kind of "reading."*

Reading is only a pale shadow of the direct perception of Nature's meanings, as is television of its images. In engaging the text of the world, we are literally inserted within the story. The details, the communications, are multidimensional, not two-dimensional words on a page. They touch us at all points of contact, millions upon millions of points of contact. These places of contact range from our deep unconscious to our conscious mind, from our bodies to our souls. And the meanings within the text literally interweave with us; we are interwoven with the text of the world.

The uses of this mode of cognition to gather knowledge of plant medicines or to understand disease are only specific applications of a general mode of perception. For our ancestors these were secondary uses of a general practice. They never knew when they would encounter meaning from the world that was directly pertinent to their own lives. So they allowed meanings from the world to flow into them constantly. They kept their heart-fields extended always, and only paid attention to a specific stream of meaning when it caught their attention.

When you go deeper into this, you will learn, as they did, to keep the field of your heart extended at all times. You will *know* that the world is a text that can be read, that meaning is always coming to you. Keeping the heart-field always extended allows us to feel the touch of meaning whenever we encounter it.

The edges of the heart-field
are difficult to determine
when in wild landscapes.
It is a constantly changing identity
like a coastline
whose edges contain many bays and inlets.
It extends itself
to the farthest reaches of which it is capable
when it is reactivated
within the landscape from which it emerged.

And as we deepen our capacity for direct perception, we find that all things are aware, that all are looking at us, that all are communicating with us. And these communications of meaning go deep. They literally are communicative touches of living beings, much more than mere informational bits encoded within words.

The pronoun best used when describing the specifics of this dimension is not "what" but "who."

— HENRI CORBIN

In this meaning-filled territory, we engage in a dialogue with the livingness of the world, receive the meanings it sends to us, respond with our own meanings in turn. There is no more intimate act we can know. Engaging in it, we *know* beyond doubt that we are never alone, that we are companioned by ensouled phenomena as intelligent and real and meaningful as we are. It is literally a return to the roots of life and a reconnection with the living ecosystem from which we are expressed, as only one form among many.

For Universe is not a place but an event, not a collection of solids but an interaction of frequencies. Not a noun but a verb. And though the linear mind can examine parts of Universe through ever greater magnification, the living fabric of its truth can be experienced only with an open heart. The meanings in Universe are available to any who relocate consciousness and begin to perceive with the heart.

As encountered meanings flow into us they change us, *remake* us. The things we need to become ourselves, to fill the holes within us with which we were born, come from this added dimension of the world, this landscape of meaning. Soul-making is something that happens in the world.

That landscape has been described in a thousand different yet subtly related ways. And the experience it delivered to those who approached it correctly has also always been this experience of wholeness—of recovered personhood.

— PTOLEMY TOMPKINS

When we accept the reality of this mode of perception, begin to use it regularly in a continual, participatory interweaving, we enter a geography of meaning, a territory of spirit, of which the physical forms of the world are only one aspect. Those who have gone before us have left maps of that geography. Still, each of us must personally enter this luminous territory ourself and learn the terrain one step at a time.

We make the path by walking:
You, walker, there are no roads
Only wind trails on the sea.

— ANTONIO MACHADO

When you first begin extending your heart-field routinely and allow yourself to be sensitive to the chance encounters with meaning that always occur in wild landscapes, the meanings you come upon will, perhaps, strike you first as only general impressions. You may find, as you walk on a certain piece of land, that a mood comes over you that you cannot escape. This may come not only from the living organisms of the place, the self-organized ecosystem itself, but also from something that happened there, some history of Man.

I farm the dust of my ancestors, though the chemist's analysis may not detect it. I go forth to redeem the meadows they have become.
— HENRY DAVID THOREAU

For the historical events that have occurred before us remain in the land, interwoven with the soil, set in stone. And, if your heart-field is open, they will come into you as you walk. You may wonder at this sudden shift of mood, wonder why a deep melancholy comes on you now, and then start at the sounds of musket fire as you walk deep in the forests of Manassas. For not all the lessons found are happy ones, and it takes more than years to wipe away the bloody spots made by brothers' hands.

There is a sound in war that stays in the mind and will not let go the soul of a man. When the men around you are wounded there is a sigh, or a sob, that comes up out of them—it sounds like a soft wind on a summer's day but it has a meaning in it that a summer wind will never have. You think maybe it's your imagination but you begin to listen for it and then you realize it comes out of the wounded. It comes out of them and into you and you carry it inside you until you die and maybe it doesn't let you go even then.

Later, after the fighting is done, when the cannons have ceased their awful thunder, for a little while a silence as deep as the farthest reaches of space falls upon everything left alive. Those that fortune has allowed to live, and it was through nothing unique that they possess, nothing which better men, fallen around them, did not also possess, suffer from the silence. It falls upon them in thick blankets and for a minute stuffs their eyes and ears and all their senses with its fabric. Then, too soon, the cries and whimpers and the terrible mewling of the wounded tear the silence into fragments that remain, if at all, only in the memory. There are the calls for water, the calls for a help that can never come, the calls for the sweet mothering that young men knew in a simpler time, the prayers for death. Into the living men the cries travel and they lodge in the deep recesses and will not let them go; they will hear them for as long as they live.

Around the shattered promise of these young men who will never marry, who will never bring children into the world, who will never see the look of love in a young child's eye, who will never speak with the voice of age to the next generation, lie the remains of the land: crops that will never know the harvest, the bodies of wild things that did not flee fast enough the rage of men locked in ancient struggle, great trees whose thousand years are ground into an hour's splinters, orchards that will never know the laughter of a child's swing. All lie in disarray around the wet bundles of mothers' pride and love. Then, after awhile, there is an odor that, with the sounds, enters into the living and there leaves a smudge that no amount of washing will erase. And the graves that are hurriedly dug are simple ones and shallow.

You will find, as you walk deeper into this way of perceiving the world, as you become more and more accustomed to encountering meaning and understanding just what that meaning is, that the meanings you encounter will deepen. For the majority of these meanings come from living beings; they are no mere history encoded within the land, but moment-to-moment communications. You will begin to find older meanings, set in stone long before the human expressed itself from the bacterial.

I like to think of nature as unlimited broadcasting stations, through which God speaks to us every day, every hour and every moment of our lives, if we will only tune in and remain so.
— GEORGE WASHINGTON CARVER

In a sense, these deeper meanings are sermons. But living sermons, living teachings that are meant to inform and shape us.

There are sermons in stones, aye and mud turtles at the bottoms of pools.
— HENRY DAVID THOREAU

And to each of us come the particular ones we need for our souls to be shaped the way they are intended to be. These teachings do not speak to the linear mind, like the dusty ones found in churches, but directly to the soul.

*I feel that [Creator] talks to us through these things he has created. I
know, in my own case, that I get so much consolation and so much
information in this way, and indeed the most significant sermons that it
has ever been my privilege to learn has been embodied in just that.*
— GEORGE WASHINGTON CARVER

And our lengthy contemplation of these "sermons" leads to a direct
reshaping of our interior.

*Every flower of the field, every fiber of a plant, every particle of an
insect carries within it the impress of its Maker and can—if duly considered—read us lectures of ethics or divinity.*
— THOMAS POPE BLOUNT

For the Earth is a living place of sacred teachings. It is not two-dimensional words on a page, not a static thing, but a living, ever-flowing communication of meaning.

*If you take the Christian Bible and put it out in the wind and rain,
soon the paper on which the words are printed will be gone. Our
bible is the wind and rain.*
— SALISH ELDER

The teachings themselves, you will begin to see, are in many instances
specific. You will begin to notice that you are not encountering the majority of them simply by chance. You are being drawn to them. There is a reason that you go to this mountain and not to that one. And you will find,
upon closer examination, that the teachings you encounter contain specific communications uniquely meant for you. That, in fact, they are
teachings intended to *re-form* you.

*The Maker of me was improving me. When I detected this interfer-
ence I was profoundly moved.*
— HENRY DAVID THOREAU

These teachings come on no predictable schedule, at no set time. You
will learn in this process to be open to what may come and take the time

to go into Nature on a regular basis so that you may encounter the teachings that are meant for you.

> *I trust myself to [Nature]. She may do as she will with me.*
>
> — GOETHE

When you go into Nature, you let the field of your heart lead, moving
to those things that for some reason attract you. You may feel one day the
need to walk in mountains, or when walking in a forest be drawn to a particular stand of trees. To notice these things you must, as Thoreau commented, let yourself "see with the unworn sides of your eye." It is in
peripheral vision that these things are seen, in peripheral thoughts that
their signals come. Pointed vision is the domain of the linear mind.

> *It is as bad to study stars and clouds as flowers and stones. I must let
> my senses wander as my thoughts, my eyes see without looking.
> Carlyle said that how to observe is to look, but I say that it is rather
> to see, and the more you look the less you will observe. . . Be not pre-
> occupied with looking. Go not to the object; let it come to you.*
>
> — HENRY DAVID THOREAU

And when you feel yourself touched by meaning, know that there is
something important for you in this phenomenon that called you to it. Go
to it and sit with it and strive to hear what is there for you.

> *As I was entering the Deep Cut, the wind, which was conveying a
> message to me from heaven, dropped it on the wire of the telegraph
> which it vibrated as it passed. I instantly sat down on a stone at the
> foot of the telegraph pole, and attended to the communication.*
>
> — HENRY DAVID THOREAU

The lessons, like those in linear schools, are often complex. Many
will take years to understand. You will know the importance of the lesson by the power of its touch upon you. There will be something in it that
tells you here is a teaching especially meant for you. And so you must
wrap it up carefully in your heart cloth, anchor in your experience the
moment of first contact, and from time to time take it out again, unwrap
and contemplate it. In your contemplation, many truths will emerge.
Eventually, in time, with your ripening, the lesson itself will burst into

understanding. You will stand in stillness, at the pregnant point, and you will see the teaching in its fullness, be able to rotate it and see it along any axis of rotation.

The difficult thing is to not turn these skills you are learning into merely a method for resource extraction from the nonmaterial realm—remaining the rest of the time in a linear mode of cognition. Ultimately, this mode of perception is not just a tool, it is a way of life, a mode of being. It is a world within which you are, and always were, meant to live. It ultimately becomes

if you wish it

an abode, not a place you visit from time to time.

you give up, eventually, being a bridge and cross over

Ultimately, the use of direct perception as a mode of being, as a normal way of cognition, begins to erase mind–body dualism. You begin to directly experience that there is no higher and lower, no up and no down, no better than and less than. No hierarchy at all. You begin to transcend anthropocentrism.

> *When the individual is able to enter a world in which the two aspects of yin and yang return to their original unity, the mission of these symbols comes to an end.*
> — MASANOBU FUKUOKA

It quickly becomes clear that in this mode of cognition, through this perception of Nature, all of Nature is one unified whole, all things are unremovable parts of one thing. And like molecular groupings that self-organize, this whole is also self-organized. It displays emergent behaviors, attends to the maintenance of the whole, the integrity of each part, and no part is less important than the whole. It is clear too that at the moment of self-organization of this whole that we call the Universe, something comes into being that is far, far more than the sum of the parts.

> *I think that in a world beyond words, where language is of no consequence, "God" and "nature" are one and the same. When I say "nature is God," what I mean is that the essence of nature and the essence of God are like opposite sides of the same reality. What*

appears on the surface is the physical form of nature; God lies concealed behind nature. Unfortunately, however, when one speaks of "inner and outer" or "front and back," because people conjure up images of two relative things they are unable to see nature on the outside and God on the inside as a single entity.

— MASANOBU FUKUOKA

This entity can be directly felt, directly experienced. It has many names, but only one identity.

religions are a particular mode of representation
they are not the thing itself

This identity is the center from which all things come. And it has always been clear to those who read the text of the world, who are open to the touch of life upon them, that this Mystery is so much greater than Man that it can never be understood with the linear mind. That before its presence we are very tiny indeed.

There is no place you
are not seen.

It is no secondhand God
but the stones under your feet,
The tree leaning casual
in shadows,
the wolf motionless
in moonlight,
your own soul
standing silent in darkness
next to your unconscious self

that see you,

all of you.

In spite of your
thinking
yourself safely invisible,

these beings,
their lives,
pull,
tug,
at your tethers,
and call you back
to suckle
in leaf-dappled shadow,
at the ancient breast
that suckled humans
long before Jesus
saw light of day,
or palmed iron,
or Buddha sat,
or ate mushrooms,
or man walked
on the moon.

And once you enter fully into it, you will begin to see that there is a specific geography to it. Its signposts are not of the physical world, but are unique instead to this deeper dimensional world of meaning. You see a mountain and notice that specific communications are encoded within its form. For its shape has meaning in it, and this meaning says something about the landscape you are within.

You begin to understand that just as the electromagnetic spectrum is fractalized when life flows through it, embedding communications you can perceive, the particular folding that occurs when life flows through the land also embeds communications. And so, the ever-slow, always-continuing (un)foldings of mountains, and any land you encounter, contain meanings themselves. The shapes of mountains are transforms of communications, particular messages that the perceptive heart can understand.

Some of these meanings have nothing to do with you, for you are not traveling in that direction, are not involved in that particular community of events. Others are teachings for you about the geography of meaning within which you find yourself. And still others are directional, pointing out the path you must take.

The traveler's every stride has a different meaning, a different import
from the preceding one, in that the goal to be reached is perfectly

understood and contained in each individual step when once the correct route has been chosen.

— GOETHE

The directional signs we encounter tell us where to go, direct us on our way. But direction in this place is not like direction in the physical world. Direction here is a quality, not a quantity. Travel is sometimes very fast and other times very slow. We may labor for years to make one mile, and at other times travel leagues in just a second. Sometimes the directional signs we encounter take us deep into our interior world, to some bent place within us. And we labor at this remaking, sometimes for years, before we are able to travel any farther.

You cannot know what awaits you
until you surrender and enter that darkness.
It may be that the sun shines there,
that the grass is green,
that your family awaits you
as they have been waiting,
for years upon years
of a length your sleeping self cannot imagine.

It is in the surrender
and in the turning to meet the darkness
that freedom lies.

For there is no place so dark
that the one who loves thee
and who has been set here to help thee
has not already been
and prepared thy place.

Then one day, when you least expect it, the bent place is made straight.

The purpose of labor is to learn;
when you know it, the labor is over.

— KABIR

The quality of your interior world has changed, and you suddenly find you have traveled a greater distance. Even though you feel far from the luminous world as you do the work set before you, when you suddenly enter the world again, you find you have traveled a distance. You find you have traveled through qualities of distance, not quantities of it.

this is the systole and the diastole of the work

Some of these places of interior work are common to all travelers, some are unique to each person. The territory has only one identity, but there are many paths through it. All travelers will encounter some of the signposts; none will encounter all. But you will find that there is some familiarity to the territory you are in, to the signposts you find, as if you have traveled to a land that you both know and do not know. The territory you walk will bear a resemblance to something within you.

This earth which is spread out like a map around me is but the lining of my inmost soul exposed.
— HENRY DAVID THOREAU

You will truly find that, as Sendivogius said, "The greater part of the soul is outside the body." When you travel into the meanings of the world, you travel deeper into your own soul.

Reading the text of the world is a lifelong journey. For the language is complex and many of the passages can be understood only after we have grown enough to do so. The territory demands so many new muscles that there are places we cannot walk until we have matured. We can often see the mountaintop before we can reach it. Still, there are directional signposts, "sermons" meant to help you on your way, that you will find in the stone you see at your feet and in the plant that calls you to a forest glade.

It would imply the regeneration of mankind if they were to become elevated enough to truly worship sticks and stones.
— HENRY DAVID THOREAU

These give instructions for the journey; the distance you travel depends on how well you understand the lesson. When you come to understand it, you will see that the lesson, when it was laid down within you, began a change in you, began altering the structure of your soul, and it was your contemplation of it that took you where you were meant to go.

After you have done the work, after that burst of understanding comes, you have a jewel visible only to others who have gone this way before you, a jewel of great cost. That jewel contains a deep truth given to you from the heart of the world, one intended to bring you into yourself. And, as you understand these jewels, you weave them deep into your fabric. They literally become part of who you are. And so, in a sense, the map itself is woven into your structure at the deepest levels of your being. Simply in being yourself now, you know the way.

> *Whatever great, beautiful, or significant experiences have come our way must not be recalled again from without and captured, as it were; they must rather become part of the tissue of our inner life from the outset, creating a new and better self within us, continuing forever as active agents in our building.*
>
> — GOETHE

These gifts of understanding, these teachings, are so dear that you must cultivate a sensitivity to their touch, keep them close when you encounter them, and spend the years necessary to understand their teachings. You must make a commitment to journey to ever-deeper levels of understanding, to penetrate the more complex meanings that come to you.

This part of the learning is best not rushed. The contemplation should unfold in its own time. Each touch is filled with deep meaning, much deeper than can be perceived in a brief time. So, eventually, we learn to sit with them, to let them into us, to contemplate them. To cultivate the power to see the journey to its end.

> *This song of the waters is audible to every ear, but there is other music in these hills by no means audible to all. To hear even a few notes of it you must first live here for a long time, and you must know the speech of hills and rivers. Then on a still night, when the campfire is low and the Pleiades have climbed over rimrocks, sit quietly and listen for a wolf to howl, and think hard of everything you have seen and tried to understand.*
>
> — ALDO LEOPOLD

There is a tendency to be grandiose about this journey in the beginning, perhaps for some time. But once you gain maturity with it, once you have remade yourself so that you can live in this world of living meanings for more extended periods, the grandiosity fades. It is simply what you do.

You have time then to linger, to greet the living beings you meet as fellow inhabitants, neighbors. You engage in sharing stories of the journey, revelations of meanings that you both have found. And you begin to hear stories, stories told by the elders who were here when they happened. For the stones and trees, plants and mountains, the rivers themselves, were here long before us. Their memories are long, and they will speak and tell those who come with open hearts about how things were long ago. They tell us the stories of their youth and the shaping of the world and of humankind.

I have occasional visits in the long winter evenings, when the snow falls fast and the wind howls in the wood, from an old settler and original proprietor, who is reported to have dug Walden Pond, and stoned it, and fringed it with pine woods, who tells me stories of old time and of new eternity; and between us we manage to pass a cheerful evening with social mirth and pleasant views of things, even without apples or cider, a most wise and humorous friend, whom I love much, who keeps himself more secret than ever did Goffe or Whalley; and though he is thought to be dead, none can show where he is buried. An elderly dame, too, dwells in my neighborhood, invisible to most persons, in whose odorous herb garden I love to stroll sometimes, gathering simples and listening to her fables, for she has a genius of unequaled fertility, and her memory runs back farther than mythology, and she can tell the original of every fable, and on what fact every one is founded, for the incidents occurred when she was young. A ruddy and lusty old dame, who delights in all weathers and seasons, and is likely to outlive all her children yet.

— HENRY DAVID THOREAU

In this way of being you ultimately find yourself, a truth that comes from the depth of the world that is meant for you, the companionship of ensouled beings that are true and loving and deep, and a journey that all ecstatic humans have taken since humans have been. It is our birthright to engage in this mode of cognition, and there is no human being who cannot do it. For it is essential to our natures, encoded within us as human beings.

What I have had, you may have, what I have enjoyed you may enjoy, what I have learned you may learn, it is all free, all open, all generously bestowed in man.

— LUTHER BURBANK

There are riches here that make gold pale in their light. Love that is larger than suns. A way of life that is a mode of being, a multidimensionality of experience that makes books the shadows that they are. I invite you to enter it. It is an easy thing to do.

> *Stop*
> *Take a deep breath*
> *Look at what is right in front of you.*
> *How*
>   *does*
>   *it*
>   *feel?*

EPILOGUE

It happened that, meditating on things as they are, my mind becoming sublime and my senses calmed, as one sated with pleasures or exhausted from fatigue, a being, of vast and boundless form, appeared before me and called my name and asked:
"What is it you want to know? What is it you want to see?"
— HERMES TRISMESGISTUS

THERE CAME A TIME WHEN I WAS DEEPLY TROUBLED by things that are occurring in the world. So I traveled high, to the heights of the Earth, to seek the wisdom that only mountains know.

I followed the path set before me, my body adapting itself to the rhythms of the land, to the rise and fall of the soul. In time I came to a sheltered place where I could see the sweep of the world before me, feel vast Mountain Spirit around me.

I sat and nestled myself into the glade, began to breathe deeply, to relax, to let myself drop down and be held by the place. I opened my feeling wide and let the field of my heart range out far beyond me, let it touch those peaks that towered over me. Let it fill with the power of Mountain. I felt them around me then, breathed them in, let their touch fill me. And I felt again that presence that is older than Man, felt the stir of an awareness that is as far beyond me as the stars from the Sun. It shifted itself, moved, then looked upon me as I sat at its feet.

I felt the caring and love I have for these mountains until the feelings threatened to overwhelm me. Then I felt the power of my need, made full the question I had come to ask. I sent my plea up and out into the world.

"Tell me," I implored, "why do we do these things we do, why are we so unkind to the Earth? Just what is the ecological function of the human species?"

I felt a pause and then a deep probe flashed through and into me. I felt myself examined, my deepest self penetrated, opened completely to a searching gaze.

"Is this what you want to know? Is this what you want to see?" came to me then.

"Yes," I replied, and sent again the earnest power of my need and caring up and out into the world.

A great caring flowed back to me and I felt myself held within the embrace of a being almost as ancient as Earth herself. Somewhere within it I sensed a deep laughter, too, as if it were remembering some vast, humorous thing.

"Are you sure," it said again, "that this is what you want to know?"

"Yes," I replied once more. "For without understanding this, how will any of us truly know what we are?"

"Then look," came the reply, "and listen."

Then the glade faded from my sight, and on the screen of my vision, forms began to take shape.

EXERCISES FOR REFINING THE HEART AS AN ORGAN OF PERCEPTION

*Begin reading Nature of your own accord first, then when you come
here you will learn to interpret it with great rapidity.*
— GEORGE WASHINGTON CARVER

I HAVE USED THE FOLLOWING EXERCISES with myself for more than thirty years
and with my students for more than twenty. They can help tremendously in
improving your ability to use the heart as an organ of perception, in refining
your perceptions, in developing facility with emotional communications, and
in reclaiming parts of yourself you may have put away long ago.

I recommend at least a year of weekly or daily work with these exercises to begin to habituate the skills.

EXERCISE 1: THE HUMAN WORLD

Funny how we use less sense than birds and animals.
— GEORGE WASHINGTON CARVER

Take a day or an afternoon and go to a part of your town that you like.
Choose a part of town in which you feel naturally happy, that feels good
to you. You are just going to be walking and visiting stores.

Begin by walking in the particular part of this area that you enjoy
most. Let yourself sink into the feeling of the place, become immersed in
it, relax into its nature.

Now. Look around you and pick the store you feel most drawn to. Walk to it and stand in front of it. Let yourself receive sensory impressions from it. Allow them to grow strong in your experience. Notice now the feelings that come to you from these sensory impressions. Let yourself explore them, touch their edges and shapes. Give yourself permission to be slow with this exploration, to not hurry. Allow yourself to notice any and all feelings that may arise, no matter how silly they seem.

In the beginning this may be confusing. The multisensory nature of human perception and feeling is so commonly repressed that it is often confusing, or scary, or awkward when you open up to it once more. Still, allow yourself to notice whatever you feel and—especially important—don't make any judgments about it. Just notice it.

Pay attention to the doors. To the windows. To what is in the windows. To the sign or signs. To the sidewalk in front of the store. To any plants or trees that may be growing there. How does each part feel to you? Do some parts feel better than others? Can you tell why?

Overall, what is the primary feeling the store communicates to you? Is it prosperous? Comforting? Happy? Somber? Melancholy? Spend as much time as you need to feel as if you have explored every aspect of the store with your feelings and come to a conclusion about it. Write everything down in a special journal that you keep for these explorations.

All this is perfectly distinct to an observant eye, and yet could easily pass unnoticed by most.

— HENRY DAVID THOREAU

Now. Look around the street. Pick another store but this time pick one that feels significantly different from the first. Go to it and repeat the process.

Compare the two stores. What different kinds of feelings did they generate? Can you tell why? Can you put this into words? (This may take some practice.)

Now go to a third store and repeat the process again. Compare your experience with the two you explored before.

Begin now to study the little things in your own door yard, going from the known to the nearest related unknown.

— GEORGE WASHINGTON CARVER

All of us unconsciously choose to go to stores or restaurants that meet emotional desires we have, to places in which we feel most comfortable, even though many other stores may sell the same things. This exercise is a process of beginning to consciously perceive and identify the embedded communications that come from the world around you and are felt in subtle emotions.

The businesses that people create embody the basic world perspectives, the underlying beliefs and orientations, that their owners possess. Businesses convey to customers specific meanings through the feelings the customers experience, though they may not normally be able to say what those feelings are. It is possible, after much practice, to identify these feelings, and from them to determine the organizational structure of a business, its level of psychological health, the impact it has on its customers, its degree of financial health, and a great many other things.

EXERCISE 2: PEOPLE

I found myself in a schoolroom where I could not fail to see and hear things worth seeing and hearing, where I could not help getting my lesson, for my lesson came to me.

— HENRY DAVID THOREAU

Now go to a coffee house that you like—one with a bookstore is a good one for this exercise—a place you can linger for a while and have some coffee or tea. Choose a place you especially like. Take a table that has a good view of the room, one where you can get a good look at the people entering the shop.

Now. Let your eye go to whichever person you are drawn to most. Really let yourself see this person. Take the time to really let his sensory impressions enter you.

Since you will be looking at him with some intensity, you will have to be clever so that you do not make him nervous or make him wonder what you are doing or why. This works best if you can observe while you yourself remain unobserved.

What kinds of feelings do you get from this person? Happy? Sad? Nervous? Empty? Masculine? Feminine? Strong? Weak? Comfortable? Assured? Indulgent wealth? Indulgent emotion? Poverty?

What thoughts come to you when you look at this person's face? Let yourself examine this face that you see. How does the chin feel to you? The nose? What is communicated from this person's eyes? Ears? Nose? Chin? Forehead? Face?

Faces are extraordinarily faithful to the internal world of their owners, no matter how schooled someone is in "keeping face." Each part of the face, through the feelings you feel, will tell you something about that person's internal world.

Now. Look at this person's hands. Do the hands seem alive and aware or asleep and unlived in? Are these hands strong or weak, happy or sad? Businesslike or filled with feeling? How old do you think this person is emotionally? Just let a number come. (Have you known other people who seem to be the same age? Are their hands similar to this person's?)

Write everything down in your journal.

How do this person's clothes feel? What do they communicate? How about the shoes? Is this person comfortable in these clothes, in the artificial skin that you see? Do the clothes match the feeling you have from looking at this person's face?

Repeat this process with as many people as you wish—at least two. Compare the experiences you had with each.

Imagine if you can a diamond made of sensitized plates like those used in the finest camera, and then conceive of the infinite variety of pictures that are printed every day—every hour!—on the plastic and impressionable mind of the child! You think that he does not see that quick, angry gesture, or hear that sharp ugly word, or feel the impatience in that push you gave him, or understand that nasty allusion, or pick up that slovenly habit; but you are wrong. All the pictures are there. Every time the lens clicks there is a permanent record.

— LUTHER BURBANK

The personal world—and the meanings—within which a person lives are communicated in every gesture, intonation, movement of eye and hand, every piece of clothing and stride of foot. It is possible, with practice, to learn to perceive all the elements of a person's internal world and their meanings, to know what it is like to live there. To understand how other people experience this person in his or her daily life. To understand the emotional tenor of the life this person lives within.

EXERCISE 3: THE NATURAL WORLD

Go and live in it. . . fish in its streams, hunt in its forests, gather fuel from its water, its woods, cultivate the ground and pluck the wild fruits. This will be the surest and speediest way to those perceptions you covet.

— HENRY DAVID THOREAU

Go to a place in Nature that you like. (Be sure and take a journal with you.) Choose a place you have been before, one with which you have some familiarity. Find the particular part of this place that you like most and let yourself relax into it. Sit down, really let yourself get comfortable.

How does this place feel? Try to describe it in words. Be as specific as you can. Go on in your journal at length if you need to. Write down everything that comes to you, no matter how silly it sounds. Even if you think it's crazy.

When you are done, allow your eyes to rove, to be drawn to whatever thing in this place that is most interesting to you. Perhaps it is a rock, a plant, or a tree.

Look at it. Let your eye explore it. Notice everything about it. Look closely at the colors, the shape, how it rests on or grows in the ground. See its relation to the air around it, to the plants, water, soil, and rocks.

Now, notice what feelings you have about this thing and the parts of it you have noticed. Let them grow very strong within you. Write all of them down.

Is there any part of what you are looking at that you like more? That you like less? Can you determine why? Do all the parts of what you are looking at generate the same feeling or emotion? Or do they generate different emotions? Write everything down in your journal in detail.

Do this with at least two other things that you see. Make sure that one of them is a plant. You can get up close if you want to, place your eye on a level with its leaf, take an insect's view of its plain. How is the plant shaped, how does it feel to your fingers, how does it smell? What emotions do each of these things generate in you? Write everything down.

The proper way is to acquire a learning directly from nature that requires no formal studies.

— MASANOBU FUKUOKA

Now, go to another natural place, different from the first. Sit down and relax. Get comfortable. How does this place feel?

Does this second place feel different from the first place? How are the feelings different? Which place feels better? Is there a name you can give the feeling you had at the first place? A name you can give the second? Names that will make clear the difference in feeling that you perceive? If you can't think of a word, make something up.

> I educated myself in my own way without adopting any given or traditional approach. This allowed me to take up every new discovery with enthusiasm and pursue the investigation of things I myself had come upon. I profited from the useful without having to bother with the odious.
>
> — GOETHE

When you are finished with this, find something else your eye is drawn to and write down everything that you feel and perceive. Do this as well with two other things, at least one of them a plant.

EXERCISE 4: THE CHILD

> All life, and particularly all animal life, is sensitive to outside impressions, but the child is far and away the most sensitive organism on the entire planet. . . The child is like a diamond, I have said more than once; its many facets receive impressions as clear and sharp as etchings.
>
> — LUTHER BURBANK

Sometimes it is helpful to make a tape recording of this next exercise and then to play it back. Instead of following the gender pronouns that I use in the exercise, use the correct pronoun for whichever gender you are. If you practice, you will find the perfect speed, pitch, and intonation for yourself to listen to.

Sit someplace comfortable. Someplace you won't be disturbed. Someplace you feel safe and nurtured.

Close your eyes and take some deep breaths. Fill up your lungs as if they were balloons, fill them to bursting. Hold it, hold it, hold it. Then . . . slowly . . . release. As you let out the air in your lungs, let any tension you feel inside you release and flow out with your breath. Do this again . . . several times.

Begin with your toes, then your ankles, your knees. As you exhale each breath, let the tension in this part of your body flow out. Do this with each major part of your body, ending with your neck, face, and head.

Now. Imagine the floor or chair under you as two huge, cupped hands holding you. Let yourself relax into them, be held by them. There is no need to hold yourself up; let yourself be supported. Keep breathing and releasing any tension in your body.

Now. See, standing in front of you, the little child that you once were. What is the impact on you of seeing this part of yourself?

Notice everything about your child. How is he dressed? How does his face look? Happy? Sad? Are you happy to see him? Does he seem happy to see you? Will he look you in the eye? Do you feel comfortable seeing him?

Notice everything about your child.

Now. Just inside yourself, ask your child if there is anything he wishes to tell you. Listen carefully to make sure you hear what he says.

Now. Is there anything you wish to tell your child?

Talk and listen as much as necessary until everything has been said. Is there anything your child needs from you? Is there anything you need from your child?

The birds seen at first by the children were sacred birds, a harmonious beauty of truth, virtue and beauty.

— MASANOBU FUKUOKA

Now. Just inside yourself, ask your child if she will give you a hug. If the answer is yes, hold your arms out in front of you (actually do this), pick your child up and bring her to you and hold her tight. Let your arms go around yourself and hold yourself tightly.

Let yourself feel this hug. Relax into it. Feel what it is like to hold this part of yourself so closely. Has it been too long since you gave yourself this kind of caring?

Now. Is there anything else you need to say to your child? Anything your child needs to say to you?

Let yourself be with this experience as long as you need or want to. Then, when you are finished, thank your child for hugging you and for coming to see you. It is always important to honor your child for helping you. Then, when you are ready, for now, say goodbye.

We give a good deal of attention to the wonder of the growth of the mind of a child, but it seems to me that the wonder does not cease with childhood.

— LUTHER BURBANK

In the Western world, especially in the United States, we are often taught to repress this part of us. It is a part of ourselves that feels very deeply and is very sensitive to the emotional nuances in the world. Reclaiming it from the bag in which it has been for so long is essential to this work. For there is no part of us that is more accessible to the field of the heart, no part that has a greater capacity to feel deeply.

Many people can have difficulty reclaiming this part of themselves. If you imagine a close friend whom you failed to meet three or four times in a row for a lunch date, you can imagine the kinds of feelings that might exist in a part of you closeted away for fifteen or twenty years. Sometimes it takes a great deal of work to reestablish communication, even more to reestablish trust. This part does not respond well to demands or threats, but will often respond to promises, especially if they are kept. (Usually you will have to do something in exchange. It is very important that you do it if you agree to.) It is worth the work it takes to make friends with this part of you again.

Opening the door to this part of you opens the door to reconnection to the world and all the subtle meanings within it. I often suggest that people do this exercise daily for at least a year. This part of yourself will tell you everything that is going on inside you, everything you deeply need. It will also tell you much about the world around you. It truly is possible to become your own best friend.

It is not necessary to tell a child, "This is wood sorrel. It looks like clover, but it's not." A child does not understand and has no need for botanical knowledge. Teach a child that clover is a green manure plant and that pearlwort is a medicinal herb useful for treating diabetes and the child will lose sight of the true reason for that plant's existence. All plants grow and exist for a reason. When we tie a child down with petty, microcosmic scientific knowledge he loses the freedom to acquire with his own hands macrocosmic wisdom. If children are allowed to play freely in a world that transcends science, they will develop natural methods of farming by themselves.

— MASANOBU FUKUOKA

There is a reason why Luther Burbank, George Washington Carver, Helen Keller, and a great many indigenous plant peoples were all said to be like children.

EXERCISE 5: THE INFANT

The integrity of a structure is compromised and perhaps made unsafe, if any portion is degraded or removed. It is the same with a person or ecosystem. The health of people or places increases with the diversity of their expression..

— JESSE WOLF HARDIN

Sometimes it is helpful to make a tape recording of this exercise—just as you did the last one—and play it back. Instead of following the gender pronouns that I use in the exercise, use the correct pronoun for whichever gender you are. If you practice you will find the perfect speed, pitch, and intonation for yourself to listen to.

Sit someplace comfortable. Someplace you won't be disturbed. Someplace you feel safe and nurtured.

Close your eyes and take some deep breaths. Fill up your lungs as if they were balloons, fill them to bursting. Hold it, hold it, hold it. Then . . . slowly . . . release. As you let out the air in your lungs, let any tension you feel inside you release and flow out with your breath. Do this again . . . several times.

Begin with your toes, then your ankles, your knees. As you exhale each breath let the tension in this part of your body flow out. Do this with each major part of your body, ending with your neck, face, and head.

Now. Imagine the floor or chair under you as two huge, cupped hands holding you. Let yourself relax into them, be held by them. There is no need to hold yourself up; let yourself be supported. Keep breathing and letting any tension in your body go.

See, lying on the floor in front of you, the little baby that you once were. What is the impact on you of seeing this part of you?

Notice everything about the baby. How is she dressed? How does her face look? Happy? Sad? Are her eyes open? Or closed?

Are you happy to see her? Do you feel comfortable seeing her? Does she seem happy to you? Is your baby moving? Is your baby breathing? What is the color of your baby's skin? Does your baby seem healthy? Or unhealthy? Is she getting enough food to eat?

Notice everything about your baby.

keep breathing

Now. When you are done noticing everything about your baby, reach down (really do this) and pick your baby up. Hold her to your chest as you would hold and cuddle any baby, let her nestle in. Feel what it is like to hold this part of yourself so closely.

Now. Even if you are a man, begin breast-feeding your baby. Allow the food from inside you to flow out and into this most vulnerable part of you. Is a nurturing happening now that has been too long absent? How long has it been since you comforted and took care of this most vulnerable part of yourself?

How do you feel doing this?

Now. As you are holding and feeding your baby, notice: is there anything your baby needs from you? Anything it wants you to do? And, as well, is there anything you need from your baby?

In a little while it will be time to stop. But before you do, is there anything you need to say to your baby? Is there anything else your baby needs from you?

Let yourself be with this experience as long as you want to. Then, when you are finished, look at your baby, allow the caring inside you to flow out and into her until she is filled up with it. And, when you are ready, thank your baby for coming to be with you, and, for now, say goodbye.

How can we expect to understand Nature unless we accept like children these her smallest gifts?

— HENRY DAVID THOREAU

This tiny, vulnerable part of us is one that is often put in the bag of shadow. It is a part of us that is helpless and needs a special kind of food. This part of us is also very important, for it knows how to suckle at the breast of the world, to take that food into itself. And this part of you is very very sensitive to emotional fields and their communications. For this part of you is the one that developed within the electromagnetic field of your mother's heart. And it knows those fields as intimately as it knows anything.

*There [in Nature] I can walk, and recover the lost child that I am
without any ringing of a bell.*

— Henry David Thoreau

Infants have no words, as you might have discovered—they perceive
in feeling-gestalts—but that is all right, the child you met first knows lots
of words. And, if you ask for his or her help, that older child is often willing to act as an interpreter.

You can repeat this exercise, if you wish, with any developmental
age you have lived through, from infancy to two, to four, to eight, to
adolescence, young adulthood, middle age, and so on. Each has its own
intelligence, its own special connection with the world. Developmental
stages do not stop at twelve or sixteen; the child naturally grows to
forty. . . and to eighty. It is possible to remain filled with feeling and
wonder and openness at any age. Each age has its own teachings. Each
is a unique developmental stage of a human being's growth. Each brings
special perceptions and capacities that aid in the experience of the
human condition.

Exercise 6: The Body

*[The followers of science] have failed to restore to the human spirit
its ancient right to come face to face with Nature.*

— Goethe

Sit someplace comfortable. Someplace you won't be disturbed. Someplace
you feel safe and nurtured.

Close your eyes and take some deep breaths. Fill up your lungs as if they
were balloons, fill them to bursting. Hold it, hold it, hold it. Then . . . slowly
. . . release. As you let out the air in your lungs, let any tension you feel inside
you release and flow out with your breath. Do this again . . . several times.

Begin with your toes, then your ankles, your knees. With each exhale
of breath, let the tension in this part of your body flow out. Do this with
each major part of your body, ending with your neck, face, and head.

Now. Imagine the floor or chair under you as two huge, cupped hands
holding you. Let yourself relax into them, be held by them. There is no
need to hold yourself up; let yourself be supported. Keep breathing and
letting any tension in your body go.

Now. See standing in front of you your lungs. Notice their shape, their color. How healthy do they seem to you? Do they seem happy? Or sad? Mad? Or scared?

Let your gaze focus on them until they come truly alive in front of you. What feelings do you have about your lungs? Let these feelings grow in intensity until they are all that you feel.

What part of your lungs stands out most clearly to your gaze? What is it about this part that is demanding your attention? What does it need from you?

Is there anything your lungs need from you? Anything you need from your lungs?

Do this until you can look at your lungs without any discomfort.

Now. Repeat this with your heart, your gastrointestinal tract, and your skin—or with any organ that seems to need your attention.

Do this until you can look clearly at any organ within you and see it from multiple points of view. Do this until you have established a communication with all your organ systems, until you feel comfortable with each of them.

EXERCISE 7: ORGAN SYSTEMS

The fragrance of that knowledge! It penetrates our thick bodies, it goes through walls—

— KABIR

Go someplace public in which you can relax and stay for a while, someplace you like. Someplace where lots of people are coming and going. An outdoor mall is a good choice. (Indoor malls are terrible for this because of the feeling-tones they generally possess).

When you are settled, let yourself relax and begin to look at the people around you. Let your eyes be drawn to whichever person most catches your attention. Now. What part of his body are you most drawn to? Let yourself focus on this part of the body and, when you are ready, let your gaze drop down below this part into the organ system underneath.

How does it look to you? Healthy or unhealthy? How does it feel: mad, sad, glad, or scared? Let yourself practice seeing deeper in this way with as many people as you want to.

Repeat this often, until you can drop down into organ systems in other people more easily.

EXERCISE 8: GOING DEEPER INTO THE HUMAN WORLD

How few are those who feel themselves inspired by what is really visible to the spirit alone!

— GOETHE

Repeat exercise 1. Go to the same places. However, ask your child to be present with you this time, perhaps standing beside you and invisibly holding your hand. Let yourself relax and really begin to see and feel the store you are looking at once again. How does it feel to you today? Remember everything you know about it.

Now. Ask your child how he or she feels about this store. What part of the store feels best to your child? What part does she like most? Ask your child to tell you everything she feels and notices about the store. Spend as much time as your child needs in order to hear everything that she has to say. Are there any differences from when you went alone to the store? What are they?

Now. Ask your infant to come. Hold your infant in your arms. How does your infant feel about this store? Notice everything that your infant does in the presence of this place. Ask your child what your infant is feeling if you have difficulty figuring it out.

Go to at least one more place that you went the first time you did this exercise and repeat this with both your child and infant. What are your child's feelings and perceptions? What are your infant's feelings and perceptions? Which place do they like better? Why? When you are ready to stop, make sure that before you do, you thank your child and infant for helping you.

EXERCISE 9: GOING DEEPER INTO PEOPLE

Not by constraint or severity shall you have access to true wisdom, but by abandonment, and childlike mirthfulness.

— HENRY DAVID THOREAU

Now, repeat exercise 2. Take your child with you again. Once you are settled, ask your infant to join you. How do they feel about this place? Do they like it or not? Why or why not? Does it feel different to you with them here?

When you are settled and comfortable, begin looking at people again. Pick the one that your child is most interested in. Have him tell you everything he perceives about that person. Does he like the person or not? Why or why not? Notice, too, once your child is finished, how your infant seems in the presence of the person. How does your infant feel about this person?

When you are done, have your child pick someone else. If he is willing, have him choose someone he is uncomfortable with. Ask your child to tell you why. What is it about that person that is uncomfortable? Have your child go into as much detail as possible.

EXERCISE 10: GOING DEEPER INTO NATURE

To know how cherries and strawberries taste, ask children and birds.
— GOETHE

Repeat exercise 3. Go to the same place in Nature you went before. Remember to take a journal. Find and sit in the same place. Let yourself relax. Imagine the Earth upon which you are sitting as huge hands holding and supporting you. Take some deep breaths.

Now. Ask your child to come and sit with you. Have your child tell you everything about this place. Go to the plant you sat with before. Touch it, smell it. Have your child touch and smell it. Have him or her tell you everything he or she knows about it. Write it all down.

Whoever follows [Nature] trustingly she takes to her heart like a child.
— GOETHE

Now. Ask your infant to come and lie on the grass in front of you. How does your infant seem today? Notice everything this part of you does in relation to the plant. Write everything down. If you want ask, ask your child to tell you how the infant feels about this plant and what your infant thinks about it.

Now. Let your child choose another plant—one your child really feels drawn to. Have him or her tell you everything about it. Write it all down. And, once again, ask your infant to come, and notice everything your infant does in response to this plant.

When you are done, leave this first place and repeat the process at the second place you went to last time you did the exercise. When you are done, before you leave, make sure you thank both your infant and child for helping.

I suspect that the child plucks its first flower with an insight into its beauty and significance which the subsequent botanist never retains.
— HENRY DAVID THOREAU

Sometimes it helps to later go and look up the plants in a book, perhaps a medicinal herb guide. The depth of information that the child and infant can gather from plants is truly amazing. In accessing these parts of yourself, you access their natural sensitivity to the world. And even if you think you are very sensitive already, you will be surprised at how much more information will come through these exercises.

People who have done these exercises with me over the years have described in detail information about plants they did not know and had never seen before. They have related medicinal, craft, clothing, and building uses that are exceptionally sophisticated and are not apparent from the exterior appearance of the plant. I have even put plants in a closed box and heard a person's child describe them in detail when the person was unable to do so earlier. It seems amazing, but it is not. It is just the way things are.

Great secrets still lie hidden; much I know and of much I have an intimation.
— GOETHE

BIBLIOGRAPHY
WITH COMMENTARY

THE WISDOM OF THE
EARTH POETS

I had long been convinced that there was nothing new under the sun and that one can find hints in transmitted knowledge of what we ourselves are just discovering, thinking about, and even producing. We are original only because we know so little.

— GOETHE

It is especially difficult for modern people to conceive that our modern, scientific age might not be an improvement over the pre-scientific period.

— MICHAEL CRICHTON

ALTHOUGH NOT ALL THE WRITERS REPRESENTED IN THIS BIBLIOGRAPHY have used direct perception many of them have entered the heart of the world and gathered information there. In so doing, they become what I call Earth Poets. The most accomplished of these are included in the Direct Perception and Miscellaneous sections. Highly recommended works are marked with a star (*).

Direct Perception

All around us we see evidences that there may be a sixth sense, of some additional power of getting impressions and knowledge from without by another means than smelling, tasting, seeing, hearing, or feeling.

— Luther Burbank

The use of direct perception in gathering knowledge from the heart of the world is extremely ancient. Published interviews with indigenous peoples show that it is pervasive throughout all cultures and all times. Less well known is that this capacity has also been described in published materials about people within the Western tradition. The following texts offer a sample from both indigenous and Western traditions.

Indigenous

A man once went to the woods and remained in solitary meditation for four days. He wandered alone till he heard a soft, low sweet voice, singing a song. He listened and watched. He saw a beautiful little flower, swaying gracefully back and forth. He knew the song came from the little flower. Around the flower the ground was swept clean. He listened until he had learned the song.

— Swimmer

Indigenous peoples have always used direct perception to understand plant medicines and the healing of disease. The largest body of direct writings on their approaches are interviews conducted primarily by ethnologists and ethnobotanists in the nineteenth century. Bruce Lamb's books on Manuel Cordova Rios's training and experience in this field are essential reading.

✴ Buhner, Stephen Harrod. *The Lost Language of Plants: The Ecological Importance of Plant Medicines to Life on Earth.* White River Junction, Vt.: Chelsea Green, 2000.

✴ ———. *Sacred and Herbal Healing Beers: The Secrets of Ancient Fermentation.* Boulder, Colo.: Siris Press, 1998.

✴ ———. *Sacred Plant Medicine: Explorations in the Practice of Indigenous Herbalism.* Coeur d'Alene, Idaho: Raven Press, 2001.

Densmore, Frances. *Teton Sioux Music.* Washington, D.C.: Smithsonian Institution: Bureau of American Ethnology, Bulletin 61, 1918.

Meyerhoff, Barbara. *Peyote Hunt,* Ithaca, N.Y.: Cornell University Press, 1974.

———. "Shamanic Equilibrium: Balance and Mediation in Known and Unknown Worlds," in *Folk Medicine.* Ed.Wayland Hand. Berkeley, Calif.: University of California Press, 1976.

✶ Rios, Manuel Cordova (F. Bruce Lamb). *Rio Tigre and Beyond.* Berkeley, Calif.: North Atlantic Books, 1985.

✶ ———. *Wizard of the Upper Amazon.* Boston: Houghton Mifflin, 1974.

Paracelsus (1493–1541)

Since nothing is so secret or hidden that it cannot be revealed, everything depends on the discovery of those things which manifest the hidden.
— Paracelsus

He who wishes to explore Nature must tread her books with his feet.
— Paracelsus

Aureolus Phillippus Theophrastus Bombast, known as Paracelsus, was a physician during the sixteenth century. His works are almost unreadable now. Though parts of them are quite wonderful, the language is generally obtuse and dated. He is only one Western practitioner in a line stretching back to the ancient Greeks who used direct perception to understand the medicinal applications of plants.

✶ Goodrick-Clarke, Nicholas. *Paracelsus: Essential Readings.* Berkeley, Calif.: North Atlantic Books, 1999.

Jacobi, Jolande, ed. *Paracelsus: Selected Writings.* Princeton, N. J.: Princeton University Press, 1951.

Paracelsus. *The Hermetic and Alchemical Writings,* 2 vols. Berkeley, Calif.: Shambhala Press, 1976.

Sigerist, Henry, ed. *Paracelsus: Four Treatises.* Baltimore, Md.: Johns Hopkins University Press, 1941.

Johann Wolfgang von Goethe (1749–1832)

What I had undertaken to do was nothing less than to present to the physical eye, step by step, a detailed, graphic, orderly version of what I had previously presented to the inner eye conceptually and in words alone, and to demonstrate to the exterior senses that the seed of this concept might easily and happily develop into a botanical tree of knowledge whose branches might shade the entire world.

— GOETHE

Goethe was perhaps the most famous personage of his time. He lived during one of the most change-filled periods in history, knew and corresponded with nearly every genius of the time, met Mozart and Napoleon, read Benjamin Franklin's work on electricity when it was first printed, was wealthy and the friend of kings. He was raised with money, and his rise to fame began when a teenager. He was acclaimed nearly his whole life.

The best general work on Goethe is by Henri Bortoft. The compilation by David Seamon and Arthur Zajonc is good, though a bit repetitive. Goethe's own writings are extensive but their quality depends on the elegance of the translator. They contain much that is not included in Bortoft and Seamon. One of the better general explorations of Goethe's thought is the lengthy (and sometimes boring) work by Johann Peter Eckermann, a collection of conversations the two had in the years just prior to Goethe's death.

If not for the importance of his poetry, Goethe's work on the direct perception of Nature (and the discoveries he made with it) would most likely be forgotten in our time (as it mostly is) because of his attacks on Newton and scientific reductionism.

✶ Bortoft, Henri. *The Wholeness of Nature: Goethe's Way of Science.* Hudson, N.Y.: Lindesfarne Press (Floris), 1996.

✶ ———. *Goethe's Scientific Consciousness.* Kent, England: The Institute for Cultural Research, 1986.

Eckermann, Johann Peter. *Conversations of Goethe.* N.Y.: DeCapo Press, 1998.

von Goethe, Johann Wolfgang. *Goethe, the Collected Works,* Vol. 12. Ed. and trans. Douglas Miller. Princeton, N.J.: Princeton University Press, 1988.

————. Goethe's *Botanical Writings*. Trans. Berthe Mueller. Woodbridge, Conn.: Ox Bow Press, 1989.

Seamon, David and Arthur Zajonc, eds. *Goethe's Way of Science.* Albany, N.Y.: State University of New York Press, 1998.

Steiner, Rudolph. *Nature's Open Secret: Introduction to Goethe's Scientific Writings.* n.p.: Anthroposophic Press, 2000.

Henry David Thoreau (1817–1862)

How indispensable to a correct study of Nature is a perception of her true meaning.

— Hᴇɴʀʏ Dᴀᴠɪᴅ Tʜᴏʀᴇᴀᴜ

Thoreau's work gave a voice to the deep passion that Americans have for the North American continent, and he is considered to be America's first important naturalist. However, this is an exceptionally shallow view of his life and work. He immersed himself within the world and worked to remake himself so that he could understand the language of Nature as clearly as possible. Regrettably, he died much earlier than the other Earth poets in this book, such as Goethe and Burbank. He had been compiling a great many observations on plants just prior to his death. What he was going to do with them no one knows; that part of his work was left incomplete. Even though he wrote prodigiously, many hours each day, his handwriting is horrendously bad and is nearly unreadable. This is why it has taken so long for much of his work to appear in print and why some of the books listed here have such recent printing dates. The best overviews are Odell Shepard's *The Heart of Thoreau's Journals* and Robert Bly's *The Winged Life.*

* Bly, Robert. *The Winged Life: The Poetic Voice of Henry David Thoreau.* San Francisco: Sierra Club, 1986.

* Shepard, Odell. *The Heart of Thoreau's Journals.* N.Y.: Houghton Mifflin, 1927 (Dover edition reprint, 1961).

Thoreau, Henry David. *Faith in a Seed.* Washington, D.C.: Island Press, 1993.

————. *The Journal of Henry David Thoreau:* In Fourteen Volumes Bound as Two. N.Y.: Dover Publications, 1962.

———. *Walden and Other Writings of Henry David Thoreau.* Ed. Brooke Atkinson. N.Y.: Random House (Modern Library edition), 1937.

———. *Wild Fruits.* N.Y.: Norton, 2000.

Luther Burbank (1849–1926)

If you will look over the wise and the great and the useful you will find them down close to the ground.

— LUTHER BURBANK

Luther Burbank was born in New England, a member of the American working class, surrounded by farmers. He grew to be one of the best-known men in the world during his lifetime, as well known as Thomas Edison. Both he and his work were widely sought out; he was recognized by every important personage in the world. Kings, politicians, scientists, and authors traveled to his home in California to visit him.

Luther Burbank developed nearly every food plant that we now take for granted. His use of direct perception and his dismay with reductionist science are two of the primary reasons knowledge of him and his work disappeared so quickly. (His hatred of the school system—formalized education—and his general support for eugenics were two others.) The best general work is his *The Harvest of Years;* the most complete look at Burbank's work itself is the twelve-volume set published in 1914 by the Luther Burbank Society. All of the material on his life and work is out of print, but can be found through the Internet: www.abebooks.com is a good source.

✱ Burbank, Luther and Wilbur Hall. *The Harvest of Years.* N.Y.: Houghton Mifflin, 1927.

Burbank, Luther. *Partner of Nature.* Ed. Wilbur Hall. N.Y.: Appleton-Century, 1940.

———. *My Beliefs.* N.Y.: Avondale Press, 1927.

✱ Whitson, John, Robert John, and Henry Smith Williams. *Luther Burbank: His Methods and Discoveries and Their Practical Applications,* 12 vols. N.Y.: Luther Burbank Press (The Luther Burbank Society), 1914.

George Washington Carver (1864?–1943)

I want them to see the Great Creator in the smallest and apparently the most insignificant things about them. How I long for each one to walk and talk with the Great Creator through the things he has created.
— GEORGE WASHINGTON CARVER

Although they used similar methods and lived at similar times, Carver and Burbank never met, and neither, as far as I can determine, ever mentioned the other in his writings. This might be because their orientations were so different.

Burbank considered himself a scientist (as did Goethe), though what they meant by that is far different than what we mean by the term now. They meant more nearly what would be called a natural philosopher, which is what all scientists were once upon a time. The trend in natural philosophy toward specialization, to extreme reductionism, was present in Goethe's time, more severe in Burbank's, and in ours complete. *Reductio ad absurdum.*

Or their apparent lack of awareness of one another's work might stem from the extreme difference in their backgrounds.

Carver was black and was born a slave, suffering all the cultural and personal impacts of those conditions. He, like Burbank and Goethe, experienced the living power of plants and Nature when a small child and this shaped the rest of his life. Unlike Goethe and Burbank, he was a born-again Christian, and this orientation flows through and shapes much of his work.

He was much less prolific a writer than Goethe or Burbank, and although some of his collected sayings are in print, as far as I am aware he left no detailed writings of his methods.

Carver, George Washington. George Washington Carver: *In His Own Words*. Ed. Gary Kremer. Columbia, Mo.: University of Missouri Press, 1987.

———. *Soul and Soil*. Ed. with commentary by Maurice King. Nashville, Tenn.: The Upper Room, 1971.

Masanobu Fukuoka (1913–)

A life of small-scale farming may appear to be primitive, but in living such a life, it becomes possible to contemplate the Great Way. I believe that if one fathoms deeply one's own neighborhood and the everyday world in which he lives, the greatest of all worlds will be revealed.

— MASANOBU FUKUOKA

Born in Japan and immersed in Eastern culture, Fukuoka trained as a scientific farmer. In consequence, he was immersed in Western thought and approaches to Nature and farming as well as Eastern cultural outlooks. Though he is exceptionally well-read, I can find no evidence that he has had any contact with Goethe's work, unlike Thoreau and Burbank. Although highly influential in the sustainable farming movement, his work has only minimally caught on. Unlike permaculture, natural farming is highly resistant to reductionism; it cannot be reduced to a series of techniques since it is (almost) wholly based on direct perception and communication. The one place that its acceptance is somewhat in evidence is, not surprisingly, India. His work is difficult to find and, with the exception of *The One Straw Revolution,* is out of print and expensive. All of his work is worth reading.

✳ Fukuoka, Masanobu. *The Natural Way of Farming.* Tokyo: Japan Publications, 1985.

✳ ———. *The One Straw Revolution.* Mapusa, India: Other India Press, 1992. (reprint of the Rodale Press edition, 1978, with a new introduction by the publisher.)

✳ ———. *The Road Back to Nature: Regaining the Paradise Lost.* Tokyo: Japan Publications, 1987.

THE HEART

*Listen friend, this body is his dulcimer. He draws the strings tight,
and out of it comes the music of the inner universe.*

— KABIR

Research on the true nature of the heart has expanded tremendously in recent years, primarily through the work of the Heartmath Institute. While important, the work has tended to be anthropocentric, focusing almost exclusively on human physiology, health, and interaction. However, this groundwork has been essential. Much of it is brilliant, especially the research of Rollin McCraty and his colleagues. Still, it is a sad commentary on our times that so much time is now spent "proving" that our hearts feel and that feeling is important.

Most popular texts on the heart as an organ of perception are terrible, oversimplistic and poorly written. *The Heartmath Solution* by Doc Childre and the overviews by Joseph Chilton Pearce in his books are probably the best. The most elegant, and one that anticipated much of the Heartmath work, is probably James Hillman's *The Thought of the Heart and the Soul of the World*.

Texts

∗ Childre, Doc. *The Heartmath Solution*. N.Y.: HarperSan Francisco, 1999.

Gershon, Michael. *The Second Brain*. N.Y.: Harper, 1998. (Note: a brief look at the enteric or GI tract nervous system.)

Glass, Leon, Peper Hunter, and Andrew McCulloch, eds. *Theory of Heart: Biomechanics, Biophysics, and Nonlinear Dynamics of Cardiac Function*. N.Y.: Springer-Verlag, 1991.

∗ Hillman, James. *The Thought of the Heart and the Soul of the World*. Woodstock, Conn.: Spring Publications, 1995.

McArthur, David and Bruce McArthur. *The Intelligent Heart*. Virginia Beach, Va.: A.R.E. Press, 1997.

Miyakawa, Kiyoshi, H. P. Koepchen, and C. Polosa, eds. *Mechanism of Blood Pressure Waves*. Tokyo: Japan Scientific Societies Press, 1984.

Paddison, Sara. *The Hidden Power of the Heart*. Boulder Creek, Calif.: Planetary Publications, 1993.

Pearce, Joseph Chilton. *The Biology of Transcendence: A Blueprint of the Human Spirit.* Rochester, Vt.: Park Street Press, 2002.
————. *Evolution's End: Claiming the Potential of Our Intelligence.* N.Y.: HarperSan Francisco, 1992.
Pearsall, Paul. *The Heart's Code.* N.Y.: Broadway Books, 1998.

Articles

Armour, J. A. "Anatomy and Function of the Intrathoracic Neurons Regulating the Heart." In *Reflex Control of the Circulation.* Eds. I. H. Zucker and J. P. Gilmore, Boca Raton, Fla.: CRC press, 1991.
Bason, B., and B. Celler. "Control of the Heart Rate by External Stimuli." *Nature* 4 (1972): 279–280.
Blalock, J. E. "The Immune System as a Sensory Organ." *Journal of Immunology* 1132 (1984): 1067–1070.
Cantin, M., and J. Genest. "The Heart as an Endocrine Gland." *Scientific American* 254 (1986): 76–81.
de Quincey, C. "Entelechy: The Intelligence of the Body." *Advances in Mind Body Medicine* 18, no. 1 (2002): 41–45.
Feder, M. E. "Skin Breathing in Vertebrates." *Scientific American* 253 (1985): 126–142.
Frysinger, R.C., and R. M. Harper, "Cardiac and Respiratory Correlations with Unit Discharge in Epileptic Human Temporal Lobe" *Epilepsia* 31 no. 2. (1990): 162–171.
Goldberger, A. "Is the Normal Heartbeat Chaotic or Homeostatic?" *News in Physiological Science* 6 (1991): 87–91.
Goldberger, A., et al. "Chaos and Fractals in Human Physiology," *Scientific American* 262 (1990): 42–49.
Goldberger, A., et al. "Nonlinear Dynamics of the Heartbeat." *Physica* 17D (1985): 207–214.
Lacey, J., and B. Lacey, "Conversations Between Heart and Brain." *Bulletin of the Institute of Mental Health* March 1987.
————. "Two-way Communication Between the Heart and the Brain: Significance of Time Within the Cardiac Cycle." *American Physiologist* 33 (1978): 99–113.
Laird, J. "Strong Link Between Emotion and Memory." *Journal of Personality and Social Psychology* 42: 646–657.

Libby, W. I., et al. "Pupillary and Cardiac Activity During Visual Attention." *Psychophysiology* 10, no. 3 (1973): 270–294.

Marinelli, R., et al. "The Heart is not a Pump: A Refutation of the Pressure Propulsion Premise of Heart Function." *Frontier Perspectives* 5, no. 1(1995). See www.elib.com/Steiner/RelArtic/Marinelli/.

McCraty, R., et al. "The Effects of Emotions on Short Term Heart Rate Variability Using Power Spectrum Analysis." *American Journal of Cardiology* 76 (1995): 1089–1093.

McCraty, R., M. Atkinson, D. Romasino, et al., "The Electricity of Touch: Detection and Measurement of Cardiac Energy Exchange Between People." In *Brain and Values: Is a Biological Science of Values Possible?* Ed. K. Pibram, Mahwah, N.J.: Lawrence Erlbaum Associates, 1998, 359–379.

McCraty R., W. A. Tiller, and M. Atkinson. "Head-heart Entrainment: A Preliminary Survey." *Proceedings of the Brain-mind Applied Neurophysiology EEG Feedback Meeting.* Key West, Fla., 1996.

McCraty, R., B. Barrios-Choplin, et al. "The Impact of a New Emotional Self-Management Program on Stress, Emotions, Heart Rate Variability, DHEA, and Cortisol." *Integrative Physiological and Behavioral Science* 33, no. 2 (1998): 151–170.

McCraty, R., et al. "New Electrophysical Correlates Associated with Intentional Heart Focus." *Subtle Energies* 4 (1995): 251–268.

Rigney, D. R., A. L. Goldberger, "Nonlinear Mechanics of the Heart's Swinging During Pericardial Effusion." *American Journal of Physiology* 257 (1989): 1292–1305.

Russek, L., and G. Schwartz. "Energy Cardiology: A Dynamical Energy Systems Approach for Integrating Conventional and Alternative Medicine." *Advances: The Journal of Mind Body Health* 12, no. 4 (1996).

———. "Interpersonal Heart-brain Registration and the Perception of Parental Love: A 42 year Follow Up of the Harvard Mastery of Stress Study." *Subtle Energies* 5, no. 3 (1994): 195–208.

Schandry, R., B. Sparrer, and R. Weikunat. "From the Heart to the Brain: A Study of Heartbeat Contingent Scalp Potentials." *International Journal of Neuroscience* 30 (1986): 261–275.

Schwartz, G., and L. Russek. "Do All Dynamic Systems have Memory? Implications of the Systemic Memory Hypothesis for Science and

Society." In *Brain and Values: Behavioral Neurodynamics*. Ed. K. Pibram and J. King. Hillsdale, N.J.: Lawrence Erlbaum Associates, 1996.

Song, L., G. Schwartz, and L. Russek "Heart-focused Attention and Heart-brain Synchronization: Energetic and Physiological Mechanisms." *Alternative Therapies in Health and Medicine* 4, no. 5 (1998): 44–62.

Skerry, T. "Neurotransmitters in Bone." *Journal of Musculoskeletal and Neuronal Interactions* 2, no. 5 (2002): 401–403.

Stroink, G. "Principles of Cardiomagnetism." In *Advances in Biomagnetism*. Ed. S. J. Williamson, M. Mohe, G. Stroink, and M. Kotani. N.Y.: Plenum Press 1989, 47–57.

Telegdy, G. "The Action of ANP, BNP and Related Peptides on Motivated Behaviors." *Reviews in the Neurosciences* 5, no. 4 (1994): 309–315.

Tiller, W. A., et al. "Cardiac Coherence: A New Noninvasive Measure of Autonomic Nervous System Disorder." *Alternative Therapies* 2 (1996): 52–65.

Watkins, A. D. "Intention and Electromagnetic Activity of the Heart." *Advances* 12 (1996): 35–36.

THE MUNDUS IMAGINALIS

Henri Corbin was the first, as far as I know, to develop a body of writing in this field. Most of his work is embedded within books on Islamic religion and spirituality. His work is exceptionally good.

Corbin, Henri. "Mundus Imaginalis or the Imaginary and the Imaginal." Retrieved June 3, 2004 from: www.hermetic.com/bey/mundus_imaginalis.htm

✱ Hillman, James. *The Thought of the Heart and the Soul of the World*. Woodstock, Conn.: Spring Publications, 1995.

Tompkins, Ptolemy. "Recovering a Visionary Geography. Henry Corbin and the Missing Ingredient in Our Culture of Images." (*Lapis* magazine, reprint online) www.seriousseekers.com.

The Necessity for
Rigorous Self-Examination

Like this land, I too have lost parts of myself.

— Jesse Wolf Hardin

This field is poorly understood. Most American self-examination occurs within traditional, anthropocentric psychotherapy modalities. It is, in consequence, somewhat self-centered and self-perpetuating in nature. While all religions contain bodies of work on this subject, their languaging is generally mingled with the dualities that strongly reflect the mode of representation of the religion in which they occur. They are quite often more in the nature of propaganda than useful information. When examining the shadow, in religious texts, the concept of "the devil" or of being "bad" or "sinful" or "not right" in some fundamental manner inevitably emerges in one form or other. In truth, the shadow merely has to do with repression.

Carl Jung did a lot of work in this field, much of it good. But Robert Bly has done the most, through this one work that I suggest, and offers a somewhat better, less grandiose, and more expanded perspective than those found in religious writings.

* Berne, Eric. *Games People Play*. N.Y.: Grove Press, 1967.
* Bly, Robert. *A Little Book on the Human Shadow*. Ed. William Booth. N.Y.: Harper and Row, 1988.

The Nonlinearity
Of Nature

A great deal of research on chaos theory and nonlinearity is occurring. Little of it is reaching public schools, which are, regrettably, still using a nineteenth-century curriculum, approach, and description of Nature. Most of the works in this field use a lot of mathematical modeling, because the authors are attempting to prove a point to people who have an investment in not understanding nonlinearity. This makes it harder for the general public to access the material. The best introductory work (and the most enjoyable) is probably that by Benoit Mandelbrot, the father of fractal theory. Skip his mathematical proofs; they are not necessary to enjoy the text.

Although he did not use the words chaos, fractal, or nonlinear (he liked omnidirectional better), Buckminster Fuller's work in this field is exceptional. However, he is a difficult read.

Fuller, Buckminster. *Synergetics: Explorations in the Geometry of Thinking.* N.Y.: Macmillan, 1975.

⋆ Mandelbrot, Benoit. *The Fractal Geometry of Nature.* San Francisco: W.H.Freeman and Company, 1983.

Walleczek, Jan, ed. *Self-organized Biological Dynamics and Nonlinear Control.* Cambridge, England: Cambridge University Press, 1999.

West, Bruce. *Fractal Physiology and Chaos in Medicine.* Singapore: World Scientific, 1990.

BIOELECTROMAGNETISISM AND PLANT ENERGETICS

Research in the electrophysiology of plants and the use of electromagnetism by plants for communication is extremely limited. The state of research is about where the use of chemical compounds by plants for communication and individual and ecosystem maintenance was a century ago. (Even now, in spite of considerable evidence to the contrary, the majority of researchers still denies that subtle chemical communications occur among plants and other members of ecosystems.)

Although significant research was conducted at the beginning of the twentieth century, most of it was suppressed in favor of other approaches that did not indicate intelligence or intention in plants. The instruments that scientists use to measure plant electrocommunication, even now, are not very sensitive. Only in recent years have researchers in the field been slightly able to convince skeptics that internal electromagnetic communication itself does occur in plants. That it occurs among plants and members of ecosystems (or even among people) is generally denied.

Most of the research on bioelectromagnetism has been conducted by scientists concerned about the impacts of human-generated electricity on living systems. That plants have a well-developed nervous system, that they show intention in their behavior and intelligence in their actions, and that there is actually little difference between plants and animals is exceptionally threatening to current assumptions about life on Earth and interferes considerably with research orientation and outcomes. The likelihood that human-generated electricity is interfering with the healthy functioning of living systems and that, in fact, the wide use of the electromagnetic spectrum for television, power generation, radio and so on is interfering with the subtle electromagnetic communications of life forms on Earth, thus disrupting ecosystem functioning, is a powerful disincentive for research. The bottom line of too many power interests depends on a lack of exploration in this field.

None of the general works in the field are very good. The best information, I suppose, is in section two of Roger Coghill's book. The journal *Bioelectromagnetics* probably publishes the most constant research material on this subject. Although a horribly dry read, Jagadis Bose, the Indian Nobel Prize winner, conducted the most elegant work in this area in the early nineteenth century.

Cleve Backster's work on the sensitivity of plants is provocative, closely linked to indigenous assertions about plant–human communications, and vigorously attacked.

Texts

Backster, Cleve. *Primary Perception: Biocommunication with Plants, Living Foods, and Human Cells.* Anza, Calif.: White Rose Millennium Press, 2003.

Bose, Jagadis Chandra. *Growth and Tropic Movements of Plants.* London: Longmans, Green, and Company, 1929.

———. *Irritability of Plants.* New Delhi, India: Discovery Publishing, 1999.

∗ ———. *The Nervous Mechanisms of Plants.* London: Longmans, Green, and Company, 1926.

———. *Physiology of the Ascent of Sap.* London: Longmans, Green, and Company, 1923.

———. *Plant Autographs and Their Revelations.* N.Y.: Macmillan, 1927.

———. *Plant Response as a Means of Physiological Investigation.* London: Longmans, Green, and Company, 1906.

∗ Burr, Harold Saxton. *The Fields of Life.* N.Y.: Ballantine, 1973.

∗ Coghill, Roger. *Something in the Air.* Lower Race, England: Coghill Research Laboratories, 1997.

Copson, David A. *Informational Bioelectromagnetics.* Beaverton, Oreg.: Matrix Publishers, 1982.

Ksenzhek, Octavian, and Alexander Volkov. *Plant Energetics*. N.Y.: Academic Press, 1998.

Rochchina, Victoria. *Neurotransmitters in Plant Life*. Enfield, N.H.: Science Publishers, 2001.

✷ Russell, Edward W. *Design for Destiny*. N.Y.: Ballantine, 1973.

Articles

Abe, S., and J. Takeda. "The Membrane Potential of Enzymatically Isolated *Nitella expansa* Protoplasts as Compared with Their Intact Cells," *Journal of Experimental Botany* 37 (1986): 238–252.

Abe, S., et al. "Resting Membrane Potential and Action Potential of *Nitella expansa* Protoplasts." *Plant Cell Physiology* 21 (1980): 537–546.

Davies, E. "Action Potentials as Multifunctional Signals in Plants." *Plant Cell and Environment* 10 (1987): 623–631.

Davies, E., et al. "Electrical Activity and Signal Transmission in Plants. How do Plants Know?" In *Plant Signalling, Plasma Membrane, and Change of State*. Eds. C. Penel and H. Greppin. Geneva, Switzerland: University of Geneva Press, 1991, 119–137.

Davies, E., et al. "Rapid Systemic Up-regulation of Genes After Heat-wounding and Electrical Stimulation." *Acta Physiol Plantarum* 19 (1997): 571–576.

Drinovec, L. M., et al. "The Influence of Growth Stage and Stress on Kinetics of Delayed Ultraweak Bioluminescence of *Picea abies* Seedlings." *Proceedings of the International Institute of Biophysics, Conference on Biophotons*, 1999.

Hashemi, B. B., et al. "Gravity Sensitivity of T-cell Activation: The Actin Cyto-skeleton." Life Science Research Laboratories, NASA, ASGSB 2000 Annual Meeting Abstracts, 2000.

Pickard, W. F. "A Model for the Acute Electrosensitivity of Cartilaginous Fishes," *IEEE Trans Biomed Engineering*, BME 35 (1988): 243–249.

———. "A Novel Class of Fast Electrical Events Recorded by Electrodes Implanted in Tomato Shoots." *Australian Journal of Plant Physiology* 28 (2001): 121–129.

———. "High Frequency Electrical Activity Associated with Water Stress in *Zebrina pendula*." *Plant Physiology* 80, suppl. (1986): 56.

Reina, F. G., et al. "Influence of a Stationary Magnetic Field on Water Relations in Lettuce Seeds," part two. *Bioelectromagnetics* 22, no. 8 (2001): 596–602.

Roa, R. L., and W. F. Pickard. "The Use of Membrane Electrical Noise in the Study of Characean Electrophysiology." *Journal of Experimental Botany* 27 (1976): 460–472.

Senda, M., et al. "Induction of Cell Fusion of Plant Protoplasts by Electrical Stimulation." *Plant Cell Physiology* 20 (1979): 1441–1443.

Shepherd, V.A. "Bioelectricity and the Rhythms of Plants: The Physical Research of Jagadis Chandra Bose." *Current Science* 77 (n.d.): 101–107.

Stange, B. C., et al. "ELF Magnetic Fields Increase Amino Acid Uptake into Vicia Faba L. Roots and Alter Ion Movement Across the Plasma Membrane." *Bioelectromagnetics* 23, no. 5 (2002): 347–354.

Stankovic, B., Davies, E. "Both Action Potentials and Variation Potentials Induce Proteinase Inhibitor Gene Expression in Tomato." *FEBS Lett* 390 (1996): 275–279.

———. "Intercellular Communications in Plants: Electrical Stimulation of Proteinase Inhibitor Gene Expression in Tomato." *Planta* 202 (1997): 402–406.

———. "The Wound Response in Tomato Involves Rapid Growth and Electrical Responses, Systemically Up-regulated Transcription of Proteinase Inhibitor and Calmodulin and Down-regulated Translation." *Plant and Cell Physiology* 39 (1998): 266–274.

Stankovic, B., et al. "Characterization of the Variation Potential in Sunflower." *Plant Physiology* 115 (1997): 1083–1088.

Takeda, J., et al. "Membrane Potentials of Heterotrophically Cultured Tobacco Cells." *Plant Cell Physiology* 24 (1983): 667–676.

Yano, A., et al. "Induction of Primary Root Curvature in Radish Seedlings in a Static Magnetic Field." *Bioelectromagnetics* 22, no. 3 (2001): 194–199.

Zawadzki, T., et al. "Characteristics of Action Potentials Generated Spontaneously in *Helianthus Annuus*." *Physiologia Plantarum* 93 (1995): 291–297.

Web Listings

Extensive lists of journal publications on bioelectromagnetism and plants are available at the following Web sites:

www.papimi.gr/PEMFmagneplant.htm
www.cogreslab.co.uk/magrefs.htm

MISCELLANEOUS

All of the works in this section are wonderful and well worth reading.

* Bateson, Gregory. *Mind and Nature: A Necessary Unity.* N.Y.: E. P. Dutton, 1979.
* Berry, Wendell. *Life is Miracle: An Essay Against Modern Superstition.* Washington, D.C.: Counterpoint Press, 2000.
* Bly, Robert. *The Kabir Book.* N.Y.: Harper and Row, 1977.
* ———. *News of the Universe.* San Francisco: Sierra Club Books, 1980.
* Keller, Evelyn Fox. *A Feeling for the Organism: The Life and Work of Barbara McClintock.* N.Y.: W. H. Freeman and Company, 1983. See especially chapters 7–9.
* McIntosh, Alastair. *Soil and Soul: People Versus Corporate Power.* London: Aurum Press, 2001.
* Pendell, Dale. *Living with Barbarians.* Sebastopol, Calif.: Wild Ginger Press, 1999.
* ———. *Pharmakodynamis.* San Francisco: Mercury House, 2002.
* ———. *Pharmako/poeia.* San Francisco: Mercury House, 1995.
* Walker, Alice. *Living by the Word,* N.Y.: Harcourt Brace Jovanovich, 1988.

NOTES

Introduction:
1. [Quoted in] Benoit Mandelbrot, *The Fractal Geometry of Nature* (N.Y.: W.H.Freeman and Company, 1983), 28.

Chapter One:
1. Mandelbrot, *Fractal Geometry of Nature,* 25.
2. Gregory Bateson, *Mind and Nature: A Necessary Unity* (N.Y.: E. P. Dutton, 1979), 49.
3. Mandelbrot, *Fractal Geometry of Nature,* 19.

Chapter Two:
1. Ary Goldberger, "Nonlinear Dynamics, Fractals, and Chaos Theory: Implications for Neuroautonomic Heart Rate Control in Health and Disease." In *The Autonomic Nervous System.* Eds. C. L. Bolis and J. Licinio. (Geneva: World Health Organization, 1999). (Retrieved June 3, 2004 from www.physionet.org/tutorials/ndc/)
2. Friedmann Kaiser, "External Signals and Internal Oscillation Dynamics: Principal Aspects and Response of Stimulated Rhythmic Processes." In *Self-organized Biological Dynamics and Nonlinear Control.* Ed. Jan Walleczek. (Cambridge, England: Cambridge University Press, 1999), 34.
3. Adam P. Arkin, "Signal Processing by Biochemical Reaction Networks." In *Self-organized Biological Dynamics and Nonlinear Control.* Ed. Jan Walleczek, 112.

Chapter Three:
1. Joseph Chilton Pearce, *The Biology of Transcendence* (Rochester, Vt.: Park Street Press, 2002), 55.

2. Paul C. Gailey, "Electrical Signal Detection and Noise in Systems with Long-range Coherence." In *Self-organized Biological Dynamics and Nonlinear Control.* Ed. Jan Walleczek, 147–148.

Chapter Four:
1. Rollin McCraty, et al., "The Impact of a New Emotional Self-Management Program on Stress, Emotions, Heart Rate Variability, DHEA and Cortisol," *Integrative Physiological and Behavioral Science* 33, no. 2 (1998): 165.

Chapter Five:
1. Joseph Chilton Pearce, *Evolution's End* (N.Y.: HarperSan Francisco, 1992), 60.
2. Ibid., 61.
3. Beatrice Lacy and John Lacy, "Two-way Communication Between the Heart and the Brain," *American Psychologist* 33 (1978): 99–100.
4. William Libbey, B. Lacy, J. Lacy, "Pupillary and Cardiac Activity During Visual Attention," *Psychophysiology* 10, no. 3 (1973): 291.
5. Rollin McCraty, public statement, made August 2003. Retrieved from www.danwinter.com/McCratyStmt.html
6. Ibid.
7. Rollin McCraty, Mike Atkinson, and William Tiller, "New Electrophysical Correlates Associated with Intentional Heart Focus," *Subtle Energies* 4, no. 3 (1995): 252.
8. [Quoted in:] Renee Levi, "The Sentient Heart: Messages for Life," 2001. Retrieved February 10, 2004 from ww.collectivewisdominitiative.org/papers/levi_sentient.htm
9. Rollin McCraty, Mike Atkinson, William Tiller, "New Electrophysical Correlates Associated with Intentional Heart Focus," *Subtle Energies* 4(3) 1995, 256.
10. Renee Levi, "The Sentient Heart: Messages for Life," Essay #2, 2001, 4. (Online at: www.collectivewisdominitiative.org/papers/levi_sentient.htm)
11. Joseph Chilton Pearce, *Evolution's End,* 88.
12. Rollin McCraty, et al, "The Electricity of Touch: Detection and measurement of cardiac energy exchange between people, in K. H. Pibram, ed. *Brain and Values,* (Mahwah, NJ: Lawrence Erlbaum Associates, 1998), 369.

13. Jagadis Chandra Bose, *The Nervous Mechanism of Plants* (London: Longmans Green, and Company, 1926), 218.

Chapter Six:
1. James Hillman, *The Thought of the Heart and the Soul of the World* (Woodstock, Conn: Spring Publications, 1995), 47.

INDEX

BOOKS OF RELATED INTEREST

Plant Intelligence and the Imaginal Realm
Beyond the Doors of Perception into the Dreaming of Earth
by Stephen Harrod Buhner

Sacred Plant Medicine
The Wisdom in Native American Herbalism
by Stephen Harrod Buhner

The Natural Testosterone Plan
For Sexual Health and Energy
by Stephen Harrod Buhner

Plant Spirit Healing
A Guide to Working with Plant Consciousness
by Pam Montgomery

Plant Spirit Shamanism
Traditional Techniques for Healing the Soul
by Ross Heaven and Howard G. Charing

Entering the Mind of the Tracker
Native Practices for Developing Intuitive
Consciousness and Discovering Hidden Nature
by Tamarack Song

Morphic Resonance
The Nature of Formative Causation
by Rupert Sheldrake

Plants of the Gods
Their Sacred, Healing, and Hallucinogenic Powers
*by Richard Evans Schultes, Albert Hofmann,
and Christian Rätsch*

Inner Traditions • Bear & Company
P.O. Box 388
Rochester, VT 05767
1-800-246-8648
www.InnerTraditions.com

Or contact your local bookseller

* Ring to prep, we sp prim. computer?
& website? (v.B.) & poems &

WAVE
Notice them. Greeted
me, often!

"The Spiritual Eye stands immediately
at the Center of Nature."
— Hegel.
(p. 174)

⊕ p. 175-176.

sutra says
See Beauty — go to the center —
immerse yourself in it
2. Reflect / invitation
1. Preparation — empty & unplugged / gaze.
3. gaze & be intimate.
4. 1st contact — flash of recognition / reunion.
5. Hear readings (Poem)
   See readings (Photo)
6. Thank it / carry it / integrate it.
   gather it

)(Lunar (moon light) ~~[illegible]~~

The light of the moon
is the source of the sun

The running relationship
Between men + women
day & night
SUN + MOON
gazing @ each other

I feel this moon light

on me
to shine God light
reflected —
rising + setting . . . . .

(Quiet
sustaining
of the world)

"Nature
Guru"

Earth Probes
+ Direct People

I Am A Naturalist